T0142202

Simplicity is Complex

Hermann Kopetz

Simplicity is Complex

Foundations of Cyber-Physical System Design

 Springer

Hermann Kopetz
Vienna University of Technology
Baden-Siegenfeld, Austria

ISBN 978-3-030-20413-6   ISBN 978-3-030-20411-2   (eBook)
https://doi.org/10.1007/978-3-030-20411-2

This Springer imprint is published by the registered company Springer Nature Switzerland AG
The registered company address is: Gewerbestrasse 11, 6330 Cham, Switzerland

Preface

It is the objective of this book to investigate what are the characteristics of *simple* versus *complex* systems and to find out what are the *design principles* that make a *Cyber-Physical System* understandable, simple to use, and easy to maintain. The adherence to these principles in the design of a *Cyber-Physical System* should lead to a significant reduction of the system complexity.

The targeted audiences of this book are engineers, managers, and advanced students who are involved in the design and evolution of Cyber-Physical Systems and are willing to spend some time outside the silo of their daily work in order to widen their background and learn more about the pervasive problems of *system complexity*.

My interest in the topic of complexity started when I was a manager of a process control department industry more than 40 years ago. This topic kept me hostage up to today. Forty years ago, the design of a process control system (today these systems are called *Cyber-Physical Systems*) was more of an art than an engineering endeavor. The software technology at that time was concerned primarily with *functional correctness* and did not pay much attention to the *temporal dimension* of program execution that is as important as functional correctness when a physical process must be controlled.

In the following years in academia, we tried to develop a *real-time system design methodology* where time was considered a first-order citizen. By taking time from the *problem domain* and putting it into the *solution domain*, many problems in the design of Cyber-Physical Systems can be simplified. We developed the *time-triggered protocols* and the *time-triggered architecture* that are now deployed in industry. But with an increase in the functional requirements and system size, the complexity problems appeared again in a different disguise.

A sound analysis of the *complexity problem* requires some insight in cognition, human problem-solving, and parts of philosophy. After becoming an *emeritus*, I had the time to look into these diverse fascinating fields more seriously.

After the introductory chapter, this book is divided into two parts. The first part (Chaps. 2–5) deals with the foundational issues of complexity, i.e., the conceptual, psychological, and philosophical aspects of complexity and the multi-level hierarchy of models that are required to orient oneself in a complex world. The distinction between *essential data* (e-data) and *context data* (c-data) in Chap. 3 is of particular importance when a responsive real-time communication system must be designed and helped us in the design of the time-triggered protocol (TTP).

A reader who does not want to start with the rather heavy foundational issues of the first part is advised to initially jump over this part and go directly to Chap. 6. The reader may go back to these foundational issues at a later time when some deeper degree of understanding of the complexity problem is desired.

Chapter 6 points out that Cyber-Physical Systems are fundamentally different from conventional data processing systems. Chaps. 7–10 deal with the simplification and the engineering design of understandable Cyber-Physical Systems and present a framework for the development of a safety-critical CPS.

More than 100 small examples are interwoven throughout the text to demonstrate how the abstract ideas are related to concrete situations.

The present book contains the essence of my thinking about complexity accumulated over the past 40 years. In some sense, this book presents the rationale for the design of the time-triggered architecture (TTA). The TTA provides a sound engineering framework and methodology for the development of Cyber-Physical Systems.

The ideas presented in this book have grown over many years: first in the industry and then out of the numerous discussions at the Institute of Computer Engineering at the Vienna University of Technology, at IFIP WG 10.4 meetings, and in a number of exciting European research projects (particularly the AMADEOS projects on *Cyber-Physical Systems of System*, headed by Andrea Bondavalli). At this point, I would like to thank many friends who have contributed valuable comments on the earlier drafts of this book: Bernhard Froemel, Frank Furrer, Laila Gide, Radu Grosu, Brian Randell, Klaus Schmidt, Lui Sha, Rob Siegel, Wilfried Steiner, Neeraj Suri, and Martin Törngren.

Parts of the material have been derived from some of my previous publications and have been revised in order to bring them into a coherent whole.

Baden-Siegenfeld, Austria   H. Kopetz

Contents

Chapter 1
Introduction

"That's been one of my mantras—focus and simplicity. Simple can be harder than complex: You have to work hard to get your thinking clean to make it simple. But it's worth it in the end, because once you get there, you can move mountains."

—Steve Jobs

To design a simple system requires *deep technological insight* and a *strong character*.

Deep technological insight is needed to capture the essence of the problem that must be solved and to identify all those topics that are only incidental and can be disregarded in the solution. A strong character is needed to defend the *conceptual integrity* of an evolving design against all attacks by proponents of "nice-to-have features" that complicate the design and lead to uncontrolled feature interactions.

Let us look at the description of *how to start an automotive engine* from 100 years ago (at the time of the Ford Model A) to appreciate another aspect of *Simplicity is Complex*.

Example 1.1.1 taken from [Tah99]: "The following list outlines the procedures for starting your Ford Model A with the hand crank. The specifics apply to a properly tuned engine. Some variations may be required and are discussed later.

- Set the emergency brake and be sure the shifter is in neutral.
- Retard the spark by raising the left (spark) lever to the top of its quadrant.
- Lower the throttle lever approximately three notches, or until the gas pedal lowers very slightly.
- Adjust the mixture on the dash to the setting appropriate for the conditions.
- .. . .
- .. .."

Today, a single button push starts an automotive engine.

This *simplicity of use* is achieved at the expense of an enormous *complexity of design* that is hidden behind the console of a car. The tradeoff between simplicity of use and complexity of a design is prevalent in many of today's intelligent products—although the sometimes strived for simplicity of use is not fully achieved.

© Springer Nature Switzerland AG 2019
H. Kopetz, *Simplicity is Complex*, https://doi.org/10.1007/978-3-030-20411-2_1

Whereas an automotive engine at the time of the Ford Model A was simple and did not contain any electronic controllers, an automotive engine of today is made up of the mechanical engine and a control system consisting of a network of micro-processors with sensors that measure the parameters of the environment of the mechanical engine and actuators that control the exact amount of fuel and the timing of the fuel injection. Such an integration of a controlling computer system (a *cyber system*) with a physical process is called a *Cyber-Physical System (CPS)*.

In order to effectively *use* a Cyber-Physical System (CPS), a person must under-stand the *functions* of the system that are available at the *human-machine interface*. In order to design, enhance and maintain a CPS a person must also understand the internal operation of the system. Understanding means that we can explain and predict the behavior of the system by referring to a *mental model* that describes the behavior of the system at a proper level of abstraction. If the system is simple to use, the cognitive effort to construct and execute this mental model will be small. On the other hand, if there are a number of operational modes and many user-supplied and sensor inputs in each mode, the cognitive effort required to construct and execute such a mental model can be substantial. We term a system that consists of many different interrelated parts and a multitude of functions as *complex*.

In a *Cyber-Physical System (CPS)* the *cyber space* meets the *physical space*. The physical timing parameters that are characteristic for the behavior of a physical system—a mechanical engine—pose constraints that must be dealt with by the con-trolling cyber system. The design of software for a CPS is thus more demanding than the design of software for a conventional data processing system, since in addi-tion to the mere functional requirements the specified temporal constraints and the proper handling of diverse environmental effects must be assured during the opera-tion of a CPS.

Complexity is the dominant factor that determines the mental effort—and thus the *elapsed time*—required to understand a system. At present, there is the tendency to interconnect existing CPSs—e.g., in the *Internet of Things*—in order to generate *System of Systems* which enhance the functionality of existing systems and provide entirely novel services. These increased interactions and the introduction of autono-mous intelligent agents lead to a further increase in the complexity of CPSs.

1.1   A Look Back in Time

In 1947 Bardeen, Shockley and Brattain discovered the *transistor effect* that made it possible to build a digital switch on a semiconducting substrate. For this invention, they received the Nobel Prize in Physics in 1956.

This transistor effect is at the root of the microelectronic revolution and the rise of the information technology that is transforming our society. By building net-works of transistors on an integrated circuit, we can implement ever more powerful logic functions in a single device. In the five decades since the design of the first integrated circuit around 1965 *Moore's law* characterized the progress of the semi-conductor industry:

> *The number of transistors on an integrated circuit doubles approximately every 2 years.*

The feature size in an electronic device has been reduced from about *10 μm* in 1978 to about *10 nm* in 2018. This means that the area occupied by one transistor in 1978 can house 1,000,000 transistors in an integrated circuit of 2018.

The following improvements go hand-in-hand with the feature size reduction:

- An increase in the speed of transistors
- A drastic reduction of the energy required for a switching operation—one transistor of 1978 required about the same energy to operate as 1,000,000 transistors of 2018.
- An improvement in the reliability and,
- A dramatic cost reduction, since the production cost of a transistor is pre-dominantly determined by the size of the transistor.

This tremendous energy reduction and cost reduction has been accompanied by an incredible exponential rise of the numbers of transistors produced each year, resulting in a semiconductor industry with a worldwide turnover of more than 400 billion dollars annually in 2018. The approaching end of Moore's laws, accompanied by an end to the impressive energy and cost reduction, will have a profound impact on the future of information technology.

Today it is possible to mass-produce a billion-transistor-single-chip-computer for less than ten dollars. Low-cost MEMS (micro-electronic-mechanical-systems) for the acquisition of visual (camera) and auditory (microphone) data and low-cost MEMS actuators, e.g., a piezo-electric print head, complete the hardware platforms that are available today for the implementation of large Cyber-Physical Systems.

The design of the software for such a billion-transistor-single-chip-computer can run into millions of dollars. During the past five decades, the relation of hardware cost to software cost of a computer system has changed dramatically. As a consequence, the software cost of a computer product dominates in non-mass-market applications. The software cost depends on the *complexity of software*. In a number of cases, the software of a large computer system can be simplified if more hardware resources are provided to separate the different essential functions into independent encapsulated hardware-software units. From an economic point of view, this tradeoff between hardware cost and software cost should always be kept in mind.

The wide availability of mass-produced computer hardware, enabled by Moore's law, and the identification of many applications that can be supported cost-effectively by computers, required more programmers to design and test the ever-larger software systems. The many problems with and failures of large software projects were one reason why a group of experts from industry and academia gathered in Rome at the *1968 Software Engineering Conference* to discuss the growing software crisis. At this conference it was realized that the control of the complexity of software is a topic of utmost concern that requires, for its solution, the long-term concerted efforts of academia and industry—an assessment that has not changed up to today.

To lift the level of abstraction and simplify the expression of complex algorithms, Ole-Johan Dahl published in 1966 the new object-oriented simulation language *SIMULA*. An *object* encapsulates data and the associated access methods in a compact unit. In the following years, a number of different object-oriented languages, such as Java, Python and C++ have been popularized.

A property common to all these programing languages is the separation of the *description of an algorithm* from the *machine that executes the algorithm*. As a consequence, considerations of issues related to algorithm execution time are neglected. As long as the actions in cyberspace are only concerned with the *value domain* of a computation and have no direct impact on the behavior of objects in the physical word, this neglect of the *temporal dimension* of a computation is of little concern. However, in Cyber-Physical Systems (CPS), where computers must interact directly with physical objects, the point in time when the interaction occurs is as important as the behavior in the value domain (consider the above example of fuel injection in an automotive engine).

Cyber-Physical Systems are *artifacts*, i.e., human designers produce these systems. This is in contrast to natural systems that are given to us by nature. Some of the complexities of an artifact are introduced by the inherent intricacies of the problem that must be solved, while other complexities result from the neglect of established design principles that lead to understandable designs.

From an economic point of view, the complexity of Cyber-Physical Systems is a subject of major concern to society at large. The elapsed time required to comprehend the use and the design of, and to maintain a Cyber-Physical System turns out to be a significant cost element. If we can decrease this elapsed time required to understand the internal behavior of a system and simplify the use and maintenance of the system, even by a small fraction, the economic consequences would be significant. However, simplicity is not for free—it requires effort and a tradeoff with other properties of a design.

At this point, a reader who does not want to start with the rather abstract foundational issues of complexity in this first part is advised to jump over and go directly to the start of the second part at Chap. 6. Part 2 elaborates on engineering principles that lead to the design of a simple Cyber-Physical System.

Chapter 2
Understanding and Complexity

> *"Simplicity is a great virtue but it requires hard work to achieve it and education to appreciate it. And to make matters worse: complexity sells better."*
>
> — Edsger W. Dijkstra.

2.1 Introduction

Widely used terms that are commonly employed in the domain of discourse on complexity, such as *understanding*, *complexity*, *information* and *data* are ill defined and have different shades of meaning in different contexts. We are not trying to develop generally acceptable formal definitions of these terms—this seems to be impossible. It is also impossible to avoid some circularity when using words to introduce and describe a set of related concepts—we apologize for this circularity.

The purpose of this chapter is to explain what we mean when we use one of these central terms. We use *general system concepts* that are discussed in the Annex (Chap. 11) of this book.

2.2 Concepts

The famous book *The Mythical Man Month* by Frederick Brooks [Bro75], the architect of the IBM 360 Operating System, talks extensively about the importance of the conceptual integrity of a design in order to achieve simplicity. But what is a *concept*—a central term in the domain of *cognition*? We follow the thinking of Vigotsky [Vig62]:

> *A concept is a consolidated unit of human thought. We call the totality of all concepts and their relations that are familiar to a person, the conceptual landscape of that person.*

© Springer Nature Switzerland AG 2019
H. Kopetz, *Simplicity is Complex*, https://doi.org/10.1007/978-3-030-20411-2_2

This *conceptual landscape*—Boulding calls it the *Image* [Bou61]— forms the personal knowledge base that is built up and maintained in the mind of a person over her/his lifetime. This personal knowledge base consists of concepts, relations among the concepts and mental models of reality (see also Sect. 4.3). The notion of conceptual landscape is further discussed in Sect. 2.4 under the heading of *cognitive complexity*.

We use the term *construct* to denote a *specific concept* that is employed in the scientific analysis of a phenomenon. A construct is more precise than a concept. (See also the discussion about the notion of an *entity* and a *construct* in Chap. 11).

Example 2.2.1: The mathematical concept of infinity is a construct.

One of the most important—and difficult—concepts in the *domain of complexity* is the concept of *understanding*, i.e. what does a person mean when she says *she understands* a phenomenon?

Let us start with the following definition for the term understanding:

Understanding a phenomenon means that a person has access to a mental model that provides an explanation for the phenomenon. The mental model must employ concepts already familiar to the person.

An explanation of a phenomenon in the physical world maps the things and the observed phenomena in the physical world that are at the center of attention to constructs and relations among constructs, i.e., a *mental model* (see Sect. 4.3), in the conceptual landscape. It then establishes logical or mathematical relations among the constructs that mimic the forces and dependencies of the things in the physical world.

The mental model must deal with the phenomenon at an appropriate level of abstraction—preferably by employing concepts that come from the level immediately below the phenomenon (this issue is further discussed in Chap. 4 on Modeling and Chap. 5 on Multi-Level Models). The *level of understanding* depends on the set of familiar concepts—the explanatory context—a person has already in her conceptual landscape for the construction of the mental model. In many domains there exists a consensus about what counts as an appropriate explanatory context among the people that work in this domain.

Example 2.2.2: In the domain of medicine it is assumed that a medical doctor has a profound knowledge of anatomy, physiology and pathology and is familiar with the Latin terms that are used in this domain. This knowledge forms the appropriate explanatory context in the domain of medicine.

We distinguish between two types of explanations: *causal explanation* and *teleological explanation* [Wri04]. A causal explanation links a (past) cause to a (later) effect, whereas a teleological explanation links a future goal (the *telos*) to a present action that causes an effect towards reaching the future goal. Causal and teleological explanations are thus not contradictory, but complementary.

2.3 Causal Explanation

If there exists a reliable regularity between the occurrence of an observable event A, the cause, and the subsequent occurrence of an observable event B, the effect, then we say A causes B. If event A is carried out in order to cause event B then we deal with a *causal action*. These reliable regularities are the result of a large number of different observations and experiments conducted and confirmed by many scientists in an ideal setting, i.e. a setting where no disturbing phenomena are present.

Hempel and Oppenheimer [Hem48] have presented the following well-known schema (the *HO schema*) for causal explanation and causal prediction in the exact natural sciences:

Given
Statements of the antecedent conditions.
and
general laws
then a logical deduction of the
phenomenon to be explained
is entailed.

The *antecedent conditions* can be initial conditions and/or boundary conditions that are unconstrained by the general laws.

The *general laws* can be either universally valid *natural laws* that reign over the behavior of things in the physical world or *logical laws* describing a valid relation in the domain of constructs.

A natural law is a conjecture about the behavior of a natural process that is widely accepted by the scientific community, has been carefully tested, and has—up to now—not been falsified.

A natural law, such as a physical law, must hold everywhere, no matter at what level of a multi-level hierarchy is the focus of the investigation.

Example 2.3.1: According to the HO schema, the trajectory of an artillery grenade is determined by the position and orientation of the gun, the initial momentum generated by the explosive and the *natural laws* of mechanics and gravitation.

A weaker form of causal explanation is provided if the general laws in the above schema are replaced by *established rules*. There are fundamental differences between general laws and established rules. General laws are universally valid and have a somewhat axiomatic character, while established rules about the behavior of things are derived locally from more or less careful observations. The degree of accuracy and rigor of various established rules differ substantially. A special case is the introduction of *imposed rules*, e.g., the rules of an artificial game, such as chess.

Example 2.3.2: In the domain of medicine, many explanations are based on established rules that have been derived from a multitude of observations and controlled experiments.

It thus follows that between the two extremes of *scientifically explained* and *not explained at all* there is a continuum of explanations that are more or less acceptable and are relative with respect to the general state of knowledge and the opinion of the observer at a given point in time.

Example 2.3.3: Let us refer to the widely known game of billiards to give an example for a causal explanation using the HO schema. In its simplest form, there are only two balls on a billiard table, the cue ball (a white ball) and an object ball. A player uses a cue stick to strike the cue ball with a controlled force in a precise direction such that the cue ball transfers its momentum to the object ball and the object ball rolls into the called pocket of the billiard table.

In an ideal setting the HO schema can be applied to predict the movement of the object ball. If we have a precise knowledge of the initial conditions, i.e., the strike of the cue stick and the positions of the cue ball and the object ball on the table, then the general laws of Newtonian physics determine the following effects.

Example 2.3.3 is an example of a *deterministic process*, where *determinism* is defined as follows:

"The world is governed by determinism if and only if, given a specified way things are at a time t, the way things go thereafter is fixed as a matter of natural law" [Hoe16].

In a deterministic model of a system the future course of events is determined by the initial and boundary conditions and by the natural laws that reign over the behavior.

In a real-world setting—an *open system*—the phrase *specified way things are at time t* refers to an *uncountable number of conditions*. To exclude all those conditions that are considered non-relevant in a given scenario, the phrase *ceteris paribus* (meaning *all other things being equal*) is often used.

Example 2.3.4: A recent report in a newspaper said: A traffic accident was caused by the fact that summer tires instead of winter tires where used on a wintery road. There must have been many other relevant conditions—a causal field composed of many diverse causes—for the accident to occur, e.g. an inappropriate speed of the car. However, the reporter considered—*ceteris paribus*—the fact that summer tires instead of winter tires have been used as the cause of the accident.

In a deterministic model of a system, an unexpected event in the environment or a small change in the initial conditions can have an unpredictable large effect on the evolving future states.

Example 2.3.5: Let us further elaborate on the concept of determinism by looking again at the game of billiards. The predictions about the movement of the object ball, according to the *HO Schema,* are only valid in an ideal setting. However

(i) The real world is an open system and not an isolated system that is presumed in the notion of an ideal setting (and the HO schema): The billiard table may not be absolutely flat; there may be an earthquake interfering with our game of billiards. We call uncontrollable processes in the open environment (such as the mentioned earthquake) environmental dynamics.

(ii) It is not always possible to have a *precise* knowledge of the initial conditions. There is a difference between the measured value and the true value of a physical quantity (called the *measurement error*). If we go to the extreme of quantum physics, the *Heisenberg uncertainty principle* tells us that it is impossible to measure two related properties, such as position and momentum, at the same instant. Thus, at the level of quantum mechanics, the concept of determinism—which is highly useful in the macro-world— becomes meaningless.

The interactions of the parts—whether physical or by the exchange of information— significantly augment the complexity of a phenomenon. The unpredictability of the precise interactions among many parts is another obstacle to the prediction of the future of a deterministic system:

Example 2.3.6: Now let us put two object balls on our billiard table and ask the player to indirectly hit the second object ball with the first object ball to move the second object ball into a called pocket. In an ideal setting, the formal analysis of this scenario is getting quite involved, but is still doable. If we put all fifteen object balls in the form of a triangle on the pool table and ask the player to open the game with an open break that drives the cue ball into the carefully arranged fifteen object balls such that the object balls hit each other and digress in diverse directions, a formal analysis is not possible anymore. The unavoidable measurement errors of the direction and the force of the cue stick will have such an influence on the movement and the interactions (collisions) of the object balls that a prediction of the evolving scenario is hopeless, even in an ideal setting.

In many computer science publications, a restricted version of determinism—we call it *logical determinism*—is introduced. In logical determinism there is no notion of physical time, only the order of events is considered.

The unidirectional cause-effect relation plays a prominent role in our mental and legal models of the functioning of the world in order to realize intended effects or to avoid the causes of undesired effects. To quote Pattee [Pat00, p.64 onwards]: "I believe the common everyday meaning of the concept of causation is entirely pragmatic. In other words, we use the word cause for events that might be controllable … the value of the concept of causation lies in its identification of where our power and control can be effective … . when we seek the cause of an accident, we are looking for those particular focal events over which we might have had some control. We are not interested in all those parallel subsidiary conditions that were also necessary for the accident to occur, but that we could not control".

Example 2.3.7: On August 14, 2018 the Morandi highway bridge in Genoa, Italy collapsed, killing more than 40 people. At the time of the accident it was known that the 51 year-old bridge was in a fragile state such that a small trigger event could cause the bridge to collapse. There is uncertainty about the precise trigger event that caused the collapse: a lightning stroke, heavy winds, heavy traffic or something else. But should the trigger event—*that might be outside the domain of human control*—be considered the *cause* of the collapse?

2.4 Teleological Explanation

Many human actions are not purely causal actions. In most cases a causal action is embedded in a larger *teleological action* that tries to reach a final goal, a *telos*. A teleological action can be understood, if we provide a teleological explanation. Already more than 2000 years ago Aristotle, in his metaphysics [Ari15, Metaphysics V.2] talks about the final cause that refers to what we call a teleological explanation.

In a teleological action a future goal is the reason for a present action, i.e. the envisioned future influences the present. This is in gross contradiction to the firmly entrenched view of the natural sciences, where only past events and not future events can have an effect on the present state of the world.

> **Example 2.4.1:** If we further analyze the above example from the game of billiards, we find out that the described causal action is actually part of a larger teleological action. In the above example the selection of the called pocket is the result of the goal-setting phase. The determination of the initial conditions for the subsequent causal action, i.e. the choosing of the direction of and the force to the cue stick, is the result of a planning phase. Finally, the above-described causal action takes place in the execution phase to bring the ball into the called pocket.

A teleological action consists of three phases:

(i) In a first phase, the *goal-setting phase*, the *telos* (i.e., the *goal, purpose* or *intention*) must be established.
(ii) In the second phase, the *planning phase*, a plan must be developed how the goal can be reached.
(iii) In the third phase, the *execution phase*, the plan must be executed in order to reach the established goal.

In the exact sciences, such as physics, teleological explanations are viewed with suspicion and contempt. This conflict can be resolved by looking at the differences between the inanimate world and the world of living entities. In models of the inanimate world—the kingdom of the exact sciences such as physics or chemistry—there is no room for concepts like *purpose, goal* or *intention*. Therefore, the concept of a teleological action has no place in the exact sciences.

However, in biology the situation is different. Let us look at the growth of a tree out of a seed. The future *goal state* and the *growth plan* are part of the seed (*the genotype*). The *seed controls* the *growth process* that brings about the future tree (*the phenotype*).

In the domain of human actions, the teleological schema of explanation is dominant. This schema of explanation, where the goal, purpose or intention of a human user are considered, can lead to a much better understanding about the role of a technical support system—such as the service of a computer.

> *We strongly believe that the understanding of a computer system can be substantially improved, if the purpose of a service of the system is explained in detail and, referring to this purpose, a thorough rationale for the stated requirements is given.*

We will discuss this issue in more detail in Chap. 10.

2.5 Complexity

In Wikipedia we find the following definition of *complexity*:

"Complexity is generally used to characterize something with many parts where those parts interact with each other in multiple ways".

In contrast, Dvorak adopts the following, different definition of *complexity* in a NASA study on *Flight Software Complexity* [Dvo09]:

"The study adopted a simple definition of 'complexity' as how hard something is to understand or verify".

The first definition is focused on *the properties of a something* that make it complex while the second definition highlights the *amount of human effort* required to understand the *something*.

Let us look at the different viewpoints taken by these definitions: the first definition considers *complexity* as a *property of something* while the second definition considers complexity as a *relation between something and an observer* who tries to understand the *something*—in this book we call the first type of complexity *object complexity* and the second type of complexity *cognitive complexity*. Schlindwein and Ison [Sch04] in their *epistemology of complexity* make a similar distinction and call object complexity *descriptive complexity* and cognitive complexity *perceived complexity*.

2.5.1 Complexity as a Property —Object Complexity

There is considerable interest to view *complexity* as a *property* of a *something*, particularly in the field of computer science. It is tried to find properties of a program or an algorithm that can form the basis for the calculation of a meaningful *object complexity measure*. This approach has the advantage of eliminating any subjective human element from the assessment of the complexity.

> **Example 2.5.1:** The *state space complexity* of a game, an *object complexity measure*, is the number of legal game positions reachable from the initial position of the game. The state space complexity of *CHESS* is 10^{47} and of *GO* 10^{170}[Wiki].

The field of *computational complexity science* classifies algorithms according to the computational resources (execution steps, memory) required to execute an algorithm. The deep insights gained from a computational complexity analysis of an algorithm—whether the computational resources grow polynomial or exponential with the size of the input—are very helpful when comparing the performance characteristics of different algorithms.

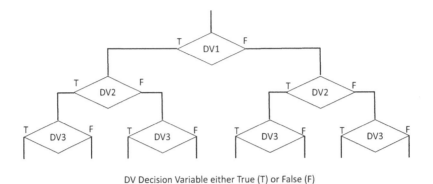

DV Decision Variable either True (T) or False (F)

Fig. 2.1 Control flow through a program with of three decision variables, resulting in a cyclomatic complexity of at most eight

In the field of computer software, the mere *statement count (lines of code LOC)*, or *McCabe's cyclomatic number* are examples of measures that try to capture the object complexity of a software component.

Example 2.5.2: If a software component contains three binary (Boolean) decision variables a total of at most $2^3 = 8$ cases must be distinguished and tested. The software component has thus a *cyclomatic complexity* of 8 (see Fig. 2.1)

The *cyclomatic number* of McCabe [MaC76] uses the number of necessary *case distinctions* of a software component, *i.e.* the number of branches in the program, as a measure of the object complexity. Since each case requires a careful analysis and an appropriate test procedure, this object complexity measure increases with the number of case distinctions.

In order to be able to exhaustively test a software component, McCabe proposes in [MaC76, p.314] that the *cyclomatic number* of a software component should be *less than ten*. Based on this proposal, we introduce the following definition of *simple* versus *complex* software:

A software component is called 'simple', if its cyclomatic number is less than ten. Otherwise the software component is called 'complex'.

Another measure for the complexity of an algorithm, expressed in a *string of characters* is the *Kolmogorov complexity*. The *Kolmogorov complexity* of a string is defined as the length of the shortest possible description of the string in some universal programming language.

2.5.2   Complexity as a Relation—Cognitive Complexity[1]

The cognitive complexity of a phenomenon assesses the *cognitive effort* required to understand a phenomenon by a human. Cognitive complexity refers to the relation between an external phenomenon and the subjective internal mental model

---

[1] This Section is a revised version of [Kop11, pp.30–36]

Table 2.1 Intuitive experiential versus analytic rational problem solving strategy (Adapted from [Eps08, p. 2])

Intuitive experiential	Rational analytic
Holistic	Analytic
Emotional (what feels good)	Logical reason oriented (what is sensible)
Unreflective associative connections	Cause and effect connections, causal chains
Outcome oriented	Process oriented
Behavior mediated by vibes from past experience	Behavior mediated by conscious appraisal of events
Encodes reality in concrete images, metaphors and narratives	Encodes reality in abstract symbols, words, and numbers
More rapid processing, immediate action	Slower processing, delayed action
Slow to change the fundamental structure: changes with repetitive or intense experience	Changes the fundamental structure more rapidly to integrate new accepted evidence
Experience processed passively and pre-consciously, seized by our emotions	Experience processed actively and consciously, in control of our thoughts
Self-evidently valid: *seeing is believing* tacit knowledge base (implicit)	Requires justification via logic and evidence explicit knowledge base

contained in the conceptual landscape of the human observer. The elapsed time needed to understand a scenario by a human of proper qualification can thus be considered a measure for the cognitive complexity.

The term *conceptual landscape* encompasses the two quite different mental sub-systems humans have available for solving problems: the *intuitive-experiential sub-system* and the *analytic-rational subsystem* [Eps08]. Table 2.1 compares some of the distinguishing characteristics of these two subsystems.

The *intuitive subsystem* is a preconscious emotion-based subsystem that operates holistically, automatically, and rapidly, and demands minimal or no cognitive resources for its routine execution. Since it is nearly effortless, it is used most of the time. It is assumed that the intuitive subsystem has access to a large knowledge base that contains a multitude of implicit mental models of the world. This subjective knowledge base, which is one part of what we call the conceptual landscape of an individual, is mainly built up and maintained by experience, drill, and emotional events that are accumulated over the lifetime of an individual.

Intuitive reasoning is holistic and has the tendency to use limited information for a quick general and broad classification of scenarios and subjects (e.g., this is a pleasant or unpleasant person). The intuitive system does integrate the data about reality in a coherent stable conceptual framework. The concepts in this framework are mostly linked by *unconscious associative connections*, where the source of an association is often unknown.

The interacting processes of *assimilation* and *accommodation* help to build the *coherent conceptual framework* [Pia70, p.63]. Whereas the simple process of assimi-lation places a new experience into the already existing structure of the conceptual landscape (the mental knowledge base), the more difficult process of accommoda-tion must modify the structure of the existing conceptual landscape to align it better with the experimental evidence. The conceptual landscape is thus continually adapted and extended, but its core structure is rather rigid and cannot be easily changed. Since a drastic change of the core structure of the conceptual landscape requires a momentous mental effort, such a change is—whenever possible—avoided.

Example 2.5.3: Face recognition—a task that a small child can perform effectively—is a good example for the power of the intuitive system. Out of many particular images of the face of a person—varying angles of observation, varying distance, changing lighting conditions—characteristic permanent features of the face are identified and stored in order that they can be used in the future to recognize the face again. This demanding abstraction process is executed unconsciously, seemingly without effort, in the intuitive subsystem. Only its results are delivered to the rational subsystem.

The rational subsystem is a symbolic reasoning system, driven by a controlled and noticeable mental effort to investigate a scenario. It is a conscious analytic subsystem that operates according to the *laws of causality and logic*. We try to get an understanding of a dynamic scenario by isolating a *primary cause*, suppressing seemingly irrelevant detail, and establishing a *unidirectional causal chain* between this primary cause and an observed effect. If cause and effect cannot be cleanly isolated, such as is the case in a feedback scenario, or if the relationship between cause and effect is *non-deterministic* then it is more difficult to understand a scenario. A typical task for the analytic-rational subsystem is the confirmation of a proof of a mathematical theorem.

There seems to be a nearly unlimited set of resources in the intuitive subsystem, whereas the cognitive resources that are available to the rational subsystem are limited [Rei10]. The seminal work of Miller [Mil56] introduced a limit of 5–7 chunks of information that can be stored in short-term memory at a given instant. Processing limitations are established by the *relational complexity theory* of Halford [Hal96]. Relational complexity is considered to correspond to the *arity* (number of arguments) of a relation.

Example 2.5.4: The relation *LARGER-THAN (elephant, mouse)* is a *binary* relation with two arguments.

The relational complexity theory states that the upper limits of adult cognition seem to be relations at the quaternary level. If a scenario requires cognitive resources that are beyond the given limits, then humans tend to apply simplification strategies to reduce the problem size and complexity in order that the problem can be tackled (possibly well, possibly inadequately) with the limited cognitive resources at hand.

There are many subtle interrelationships between the *intuitive-experiential* and the *rational* problem-solving subsystems, which form the extremes of a continuum of problem-solving strategies. In most real-life scenarios both systems cooperate to quickly arrive at a solution. Adult humans have a conscious *explicit model* of reality in their rational subsystem, in addition to their *implicit model* of reality in their intuitive subsystem. These two models of reality coincide to different degrees and form jointly the conceptual landscape of an individual. It is not infrequent that, after unsuccessful tries by the rational subsystem, at first a solution to a problem is produced unconsciously by the intuitive subsystem. Afterwards this solution is justified by analytical and logical arguments that are constructed by the rational subsystem.

Similarly, the significance of a new scenario is often recognized at first by the intuitive subsystem. At a later stage it is investigated and analyzed by the rational subsystem and rational problem-solving strategies are developed. Repeated encounters of similar problems—the accumulation of *experience*—effortful learning and drill move the problem-solving process gradually from the rational subsystem to the intuitive subsystem. The cognitive resources that have previously been allocated to the problem-solving process in the limited rational subsystem are thus released. There exist many practical examples that demonstrate this phenomenon: learning a foreign language, learning a new sport, or learning how to drive a car. It is characteristic for a *domain expert* that she has mastered this transition in her domain and mainly operates in the effortless intuitive mode, where a fast, holistic and intuitive approach to problem solving dominates.

> **Example 2.5.5:** Neuro-imaging studies have shown that the intuitive and rational subsystems are executed in two different regions of the human brain. This is illustrated by an experimental study of the chess-playing strategy of amateurs vs. grandmasters of chess. The excitations of the intuitive and the rational subsystem, observed by brain imaging studies, showed that amateurs displayed the highest activity in the medial temporal lobe of the brain, which is consistent with the explanation that their mental activity is focused on the rational analysis of the new move. The highly skilled grandmasters showed more activity in the frontal and parietal cortices, the location of the intuitive subsystem, indicating that they are retrieving stored patterns of previous games from expert memory in order to develop an understanding of the scenario [Ami01] (see also example 2.5.2 for the state space complexity of chess).

2.5.3 Complexity of a Model

We take the view of Edmonds [Edm00] that complexity can only be assigned to models of physical systems, but not to the physical systems themselves, no matter whether these physical systems are natural or artificial. A physical system has a nearly infinite number of properties—every single transistor of a billion-transistor system-on-chip consists of a huge number of dopants that are placed at distinct positions in space. We need to abstract — to build models that leave out the seemingly irrelevant detail — in order to be able to reason about the properties of interest.

> **Example 2.5.6:** If we are interested in the trajectory of a thrown stone, we abstract from most of the properties of a stone and develop a *model* that regards the stone as a mass-point with a given mass.

The perceived cognitive complexity of a model depends on the relationship between the representation of the model and the existing subjective conceptual landscape and the problem-solving capability and experience of the observer. The perceived cognitive complexity also depends on the interactions among the entities within the model. *In cyber-space the interactions are realized by the exchange of information.* The mental effort to understand a model, i.e., the cognitive complexity, increases with the number of entities in the model and the number of interactions among the entities of a model.

Example 2.5.7: If the observer is an expert, such as the chess grandmaster in the previous example 2.5.6 of chess, the intuitive subsystem provides an immediate understanding of the scenario without any real mental effort. According to our metric, the scenario will be judged as *simple*. An amateur has to go through a tedious cause-and-effect analysis of every chess-move employing the rational subsystem. This takes time and explicit cognitive effort. According to the above metric, the same chess scenario will be judged as *complex*.

There are models of phenomena that are intrinsically difficult to comprehend under any kind of representation, e.g., if the model contains many *case distinctions* as expressed by the cyclomatic number of McCabe. It may take a long time, even for an expert in the field, to gain an understanding of each case that must be considered in such a model. According to the introduced metric, these models are classified as exceedingly complex.

In order to gain a full understanding of a large system we have to understand many models that describe the system from different viewpoints at different abstraction levels. The cognitive complexity of a large system depends on the number and complexity of the different models that must be comprehended in order to understand the behavior of the complete system. The time it takes to understand all these models can be considered as a measure for the cognitive complexity of a large system.

A user who interacts with a system across a *user interface* must understand the *provided interface model*, but does not have to understand the internals of the system. For the user, the provided interface model is the system. To simplify the use of a system, this user interface model should be aligned with the conceptual landscape of the intended user and the anticipated use cases of the system. This topic is discussed extensively in Chap. 9.

2.6   Points to Remember

- A concept is a consolidated unit of human thought. We call the totality of all concepts and their relations that are familiar to a person, the conceptual landscape of that person. A construct denotes a specific concept that is used in the scientific analysis of a phenomenon.
- An explanation of a phenomenon in the physical world maps the things and the observed phenomena in the physical world that are at the center of attention to constructs and relations among constructs, i.e., a mental model, in the conceptual landscape. It then establishes logical or mathematical relations among the constructs that mimic the forces and dependencies of the things in the physical world.
- A causal explanation links a (past) cause to a (later) effect, whereas a teleological explanation links a future goal (the telos) to a present action that causes an effect towards reaching the future goal.
- A natural law is a conjecture about the behavior of a natural process that is widely accepted by the scientific community, has been carefully tested, and has—up to now—not been falsified.

- The conceptual landscape of a person comprises two subsystems, the intuitive-experiential subsystem and the analytic-rational subsystem. The intuitive-experiential subsystem is a preconscious emotion-based subsystem that operates holistically, automatically, and rapidly, and demands minimal or no cognitive resources for its routine execution. The rational subsystem is a symbolic reasoning system that operates according to the laws of causality and logic and requires a conscious mental effort.
- Understanding a phenomenon means that a person has access to a mental model in his conceptual landscape that provides an explanation of the phenomenon.
- Between the two extremes of *scientifically explained* and *not explained at all* there is a *continuum of explanations* that are more or less acceptable and are relative with respect to the general state of knowledge and the opinion of the observer at a given point in time.
- Complexity is used to characterize something with many different parts where those parts interact with each other in multiple ways. *Complexity* can only be assigned to *models of physical systems*, but not to the physical systems themselves, because a physical system has a nearly infinite number of properties.
- We distinguish between *object complexity* that is a property of a scenario and *cognitive complexity* which is a relation between a scenario and an observer.
- A user who interacts with a system across a *user interface* must understand the *provided interface model*, but does not have to understand the internals of the system. For the user, the *provided interface model is the system.*

Chapter 3
Information Versus Data

> *"You can have data without information, but you cannot have information without data."*
>
> —Keys Moran

3.1 Introduction

As outlined in the previous chapter, the number of the parts, the properties of the parts and the interactions among the parts that make up a system cause the complexity of the system. In a Cyber-Physical System (CPS) the interactions of the parts are realized by the exchange of information among the parts in cyberspace and by retrieving and sending information from/to the physical space, i.e., a user and/or a physical process. But what is *information*? What is the information content of a database?

The most often cited *theory of information* is Claude Shannon's *mathematical theory of information* which studies the fundamental limits of signal transmission and data compression. Colin Cherry in his book on *Human Communication* remarks [Che66, p.9]: "As a theory it (Shannon's mathematical theory of information*)* lies at the syntactic level of sign theory and is abstracted from the semantic and pragmatic level." A syntactic theory deals with relations among identified units without concerns about their meaning.

The focus in this book is not on the *syntactic*, but on the *semantic and temporal aspects* of information. The views on the concept *information* that are the basis for this book have been influenced by the work of Bar-Hillel and Carnap [Bar64], Kent [Ken78], and Floridi [Flo05] on *semantic information*.

It is the purpose of this chapter to explain the meaning and the use of the important terms *information* and *data* in this book and to show in Sect. 3.4 how the amount of data that must be exchanged among the components of a system can be reduced in order to improve the responsiveness of a distributed CPS. We introduce the new concept of an *information item*, called *Itom*, to capture the *meaning* (the *semantic content*) that is contained in a linguistic expression that reports about a proposition.

© Springer Nature Switzerland AG 2019
H. Kopetz, *Simplicity is Complex*, https://doi.org/10.1007/978-3-030-20411-2_3

3.2 Information

Let us start with the definition of *communication* contained in the *Merriam-Webster* dictionary:

"Communication: A process by which information is exchanged between individuals through a common system of symbols, signs or behavior; the function of pheromones in insect communication."

Information consists of one or more *information items*. By an information item we denote a proposition about a state or a sequence of states of an entity in the world at a specified time.

According to the above definition of communication, a common system of signs or symbols that are shared between the sender and the receiver must be available to realize a communication act.

A *sign* consists of three parts

- The *signifier*, i.e., a physical pattern (picture, sound, marks on a substratum, pheromone) that makes up the sign, e.g., a word written on a piece of paper,
- The *designatum*, i.e., an entity that is represented by the sign or the meaning of the sign, e.g., the concept that is designated by a word, and
- The *instant in time* when the sign is raised. If a sign is static, i.e. does not change in the interval of discourse, then there is no specific instant associated with a sign.

We distinguish between two types of signs that are used in a communication act: an *icon* and a *symbol*.

An icon is a sign where the pattern of the sign, i.e., the signifier, resembles the designatum i.e., the meaning of the sign. An example for an icon is a photograph.

Example 3.2.1: Many traffic signs are *icons* standardized by the Vienna convention of Road Sign and Signals in 1968.

If there is no obvious resemblance between the signifier and the designatum a sign is called a symbol. When dealing with a symbol, the relationship between the signifier (the pattern of the symbol) and the designatum (the meaning of the symbol) is determined by the *context,* e.g., cultural circumstances and conventions that must be learned. A word in a language is a good example for a symbol.

Example 3.2.2: Consider the whistling tone at an intersection denoting an approaching train. The whistling tone (physical pattern) is the signifier, the approaching train is the designatum and the instant in time is the point in time when the whistling tone is observed. The connection between the appearance of a whistling tone and an approaching train must be learned.

The signifier of a symbol, considered in isolation, is meaningless In order to get hold of the meaning of the signifier of a symbol, the context-dependent relationship between the signifier and the designatum must be known. We call the description of this context-dependent relationship between a signifier and the associated designatum the explanation of the symbol. The signifier is always a thing, i.e. a property of a physical object (e.g., a pattern of an object, a sound), while the designatum can be a thing or a construct.

Central to our notion of information is a new concept termed *Itom* (abbreviated from *Information Atom*). Let us start with the following definitions of an Itom:

> *Itom (Information Atom): An Itom is a semantic communication construct that informs about a proposition that holds at a specified time.*

The proposition can refer to the past, the present or the future. An Itom can inform about the existence, a state, an event, a property or a value of a property of the entity. (See the Annex for a more detailed discussion of these terms). An Itom is represented by the signifier part of a symbol. The meaning of the Itom—the designatum of the symbol— is the proposition expressed by the Itom. This proposition is revealed by the explanation of the symbol.

The meaning—the semantic content—of the Itom must be presented in a form that can be comprehended by the intended receiver of the Itom. The appropriate representation of an Itom—the physical signifier of the symbol—depends on the cultural and/or technological context of the Itom's environment, and whether the Itom is destined for a human user or for a machine. Since there exists no absolute context, the representation and the explanation of an Itom are always relative to a context, while the semantic content carried by an Itom, the information, has a connotation of absoluteness.

> **Example 3.2.3:** To demonstrate the close interrelationship between the explanation of symbols and the associated context, we elaborate on the example of the transportation of a letter between Europe and China under the assumption that the European postman can only decipher Roman Characters (RC) and the Chinese postman can only decipher Chinese Characters (CC). Which characters (signifier parts of symbols) should be used in the address fields of the envelope? Assume that an address contains the following fields: recipient's name; street name; city name and country name. A letter sent from Europe to China should use the following characters in the recipient's address field: recipient's name—CC; street name—CC; city name—CC; country name—RC). A letter sent in the other direction, from China to Europe should use: recipient's name—RC; street name—RC; city name—RC; country name—CC).

The term *context* plays an important role in our investigations. The *Oxford Dictionary* defines

> *"Context: The circumstances that form the setting of an event, statement, or idea in terms of which it can be fully understood."*

If we move an Itom to a new context, the representation of the Itom may change, but the semantic content of the Itom must remain invariant.

> **Example 3.2.4:** The Itom *Now the sun is shining*, expressed in English, has the same meaning as the Itom *Jetzt scheint die Sonne,* expressed in German, despite the fact that the two representations look fundamentally different.

3.2.1   *Properties of an Itom*

Let us now look at some of the important properties of an Itom:

Comprehensibility The semantic content of an Itom must be presented in a form that can be comprehended by the intended receiver of the Itom in the given context. In human communication, every word used in a communication act should be part of the conceptual landscape of the receiver. In machine communication, the receiving machine must contain instructions how to handle the Itom.

Purpose In human communication every Itom—as any other construct— is normally created for a purpose by a human author. In machine communication the purpose of an Itom is determined by the intent of the human designer of the machine or of the software.

Truthfulness Our concept of an Itom does not make any assumptions about the truthfulness of the proposition contained in the Itom. We thus can classify the semantic content of an Itom as true information (correspondence theory of truth [Gla13]), misinformation (accidentally false) or disinformation (intentionally false—fake news). It is often the case that only sometime after an Itom has appeared that it can be decided whether the information is true or false.

Neutrality The semantic content of an Itom does not depend on the state of knowledge of the receiver of the Itom. The aspect of newness of information to the receiver and the associated metrics about the subjective value or the subjective utility of information to the receiver is not part of our concept of an Itom.

Temporal Aspects The semantic content of an Itom is normally time-dependent. The inclusion of a timestamp that records when the Itom has been created or starts to become useful and possibly another timestamp that records when the Itom loses its temporal validity makes it possible to restrict the validity of the information to a specified real-time interval in the *Interval of Discourse (IoD)*.

Relativity While the semantic content of an Itom is independent of its representation, the representation itself is context dependent. Consider the case, when an idea is translated from one language to another language. Since there exists no absolute context, there cannot be an absolute representation of an Itom.

> **Example 3.2.5:** An example for the context-dependent representation of an Itom is an Itom that denotes the value of the proper speed of a car at a given instant. Assume that this Itom is transferred from a US context to European context. In an US context, the value is 60 and the explanation says that the number has to be interpreted as miles per hour and the timestamps informs at what instant the speed has been observed. In a European context the value is 96 and the explanation says that this number has to be interpreted as kilometers per hour. Although on the surface the two representations of the Itom are fundamentally different— with respect to the values as well as with respect to their explanations—the semantic content, i.e., the information carried by these two Itoms, is the same.

3.2.2   Human Communication

In the following Section we elaborate on the use of the new Itom concept and take as an example the communication among humans. As mentioned before, the semantic content of an Itom in the mind of the sender (the meaning of the Itom) is a proposition about a state or a sequence of states of an entity in the world. There exists no absolute representation of an Itom.

What are the steps that take place when an Itom is transmitted from one human (the sender) to another human (the receiver)? The first step in the transmission of an Itom involves a decision about the representation of the Itom in the mind of the sender. Many details have to be settled when creating the representation of an Itom, e.g., the selection of a context, the choice of a set of symbols (a specific language), the picking of the symbols (the words in the language), the place and time where and when the communication act takes place, etc.. The same semantic content—the meaning of an Itom—can be represented by very different representations.

Example 3.2.6: Consider the representation of an Itom by a sign language for the deaf-and-dumb.

We partition the totality of all details of a representation of an Itom into two disjoint sets: the *shared details* (those details common to the environment of the sender and receiver, called the *shared context*) and the *particular details* (those details that are not part of the shared context). Only the particular details of an Itom must be transmitted in the communication act from the sender to the receiver in order that the receiver can recover the meaning of the Itom under consideration.

Let us assume that the particular details of the Itom are represented in the mind of the sender in the form of a sequence of words in the chosen language. We now distinguish between two cases, written communication and oral communication.

Written communication In written communication the sender augments the sequence of words that are formed in her mind with punctuation marks (e.g., spaces, commas, etc.) to delineate syntactic units and transcribes this augmented sequence of words (signifier parts of symbols) on a physical medium, e.g. paper. At this instant, the words that originated in the human mind of the sender are embodied as a sequence of observable signifiers, a physical pattern in the physical world.

*We call the physical pattern formed by this sequence of signifiers **data**. A data item is a physical pattern that functions as the signifier part of a symbol. The associated designatum is the meaning of the data item.*

A human receiver reads the data (the sequence of signifiers) and reconstructs the particular details, i.e. words of the Itom in the mind of the reader out of the received data. By combining these particular details of the Itom (the data) with the shared details of the context that provides the explanation of the data, the receiver reconstructs the contents of the Itom under consideration.

Oral communication In oral communication the speaker encodes the sequence of words that represent the particular details of the Itom into an acoustic pattern, a *stream of sounds*. This stream of sounds, the audio data, arrives at the ears of the listener who retrieves out of this sequence the words sent by the speaker. In the next step the listener combines the particular details, i.e., the sequence of received words, with the shared details of the context to arrive at the meaning of the Itom under consideration.

Many of the processes during the transmission of an Itom from a human sender to a human receiver are taking place in the intuitive subsystem of the human mind and do not place any burden on the rational subsystem. In some situations, the choice of a right word to represent the contents of an Itom requires the support of the rational subsystem, because a right word to express the current mental experience cannot be found automatically by the speaker.

3.3 Data

In the previous Section we defined data as the physical pattern that represents the signifier of a symbol employed in a communication act.

This physical pattern can be a visual pattern, such as the letters used in written language, an acoustic pattern, such as spoken language or any other physical pattern that can be perceived by the receiver (e.g. gestures). In cyberspace, data is always a bit-pattern.

In cyberspace, a bit-pattern that represents the signifier part of a symbol is called a data item.

A *data item* considered in isolation has no meaning. The information associated with a data item (the meaning of the respective Itom) can only be grasped if an explanation of the data item is available at the receiver. If this is the case then the mere physical pattern becomes the signifier part of a symbol and the explanation establishes the link to the designatum, i.e., the meaning of the symbol.

Example 3.3.1: Let us illustrate the difference between our notions of data and information by elaborating on the well-known Rosetta Stone. When the old Egyptian Rosetta Stone with its obscure patterns (the hieroglyphic inscriptions) was discovered in 1799, the meaning of the pattern (the data) could not be explained, at least not by the scientists who found the stone. It took about 20 years until the meaning of the patterns carved in the stone was unveiled. At this instant, the hieroglyphs became symbols that disclosed information to the scientists.

We are now in the position to take a new view at the concept of an Itom:

An Itom is an atomic triple consisting of a data item, the explanation of the data item and the validity time that denotes when the proposition expressed by the Itom holds.

Fig. 3.1 Essential features
of an Itom

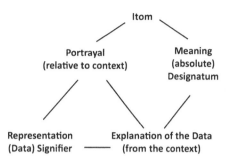

An Itom is a construct that has a meaning and is portrayed by the context dependent signifier part of a symbol. The explanation that is part of the Itom establishes the link between the signifier (the representation) i.e. the data, and the designatum, i.e., the meaning of the data contained in an Itom (Fig. 3.1).

Frequently we call the data item of an Itom also the value of an Itom. In most cases, the explanation of the data is static in the IoD (interval of discourse), while the value of the Itom, the data, is dynamic, giving rise to a meaning that changes over time. This is the reason why we include the *validity time* as part of an Itom. In case the proposition is static (i.e. it does not change during the IoD) we deal with a constant and do not need the validity time in the Itom.

Since data without an implicit or explicit explanation is meaningless, an Itom is the smallest atomic unit that can convey information.

A *data type* is an abstraction of an Itom where the value of the Itom is unspecified. A data type informs about some parts of the explanation of an Itom. When developing an algorithm, we often abstract from the concrete data and are only interested in the data type.

A *data structure* consists of a collection of related data items. Some of these data items can be dynamic while other data items of a data structure can be static to support the explanation of the data. A data structure can be expanded to represent not only one, but also two or many more Itoms in a compact form.

Example 3.3.2: Consider the case of an employee record that contains the name of an employee, the address and many other properties that are needed by an organization.

A data structure can be transported in a message from the sender to one or more receivers:

A message is a unidirectional transportation unit that captures the value domain and the temporal domain of a data-structure at a level of abstraction that is applicable in many diverse scenarios of machine communication and human communication.

The data aspect, the timing aspect, the synchronization aspect, and the publication aspect are integral parts of the message mechanism. A basic message transport service transports the data, i.e. the signifiers of the symbols, from a sender to one or a set of receivers.

3.3.1 Data for Humans

A data item destined for a human must be understandable to the intended human receiver. Understanding means that the patterns, symbols, names, concepts and relations that are used to represent the data and the explanation of the data can be linked to the existing conceptual landscape, the mind, of the intended receiver (see Sect.2.2). The understanding of the data carried by an Itom is improved if the concepts used in the explanation of an Itom are well- established in the conceptual landscape of the intended receiver. It can happen that the data carried in an Itom is incomprehensible to one receiver, while making sense to another receiver who has the proper conceptual background for understanding the presented data.

The utility of the information carried in an Itom is relative to the human who receives the information. If an Itom contains information that is new to a receiver, it causes modifications in the conceptual landscape of the receiver. The extent of these modifications is a measure of the utility of the information to this receiver (see also the difference between assimilation and accommodation of new information described in Sect. 2.5). If the receiver already knows the information contained in the Itom, no modification of the conceptual landscape is required—the information contained in the Itom has no utility to this receiver. In general, it is not possible to quantify the utility of information to a human receiver because the state of the conceptual landscape of the human receiver at the instant of receiving an Itom cannot be grasped. Bar Hillel and Carnap [Bar64] have developed a quantitative measure for the value of information based on the changes in a well-defined formal state of a hypothetical receiver— a proposal that is closely related to our view of the utility of information.

In human communication, the explanation of the data can be expressed in the form of an icon, i.e. a sign where the pattern of the sign (the signifier part of the sign) resembles the designatum (the meaning of the symbol).

Example 3.3.3: The meaning of a switch on the human-machine interface (HMI) of a globally distributed product can be explained by displaying a picture (an icon) of the object that is turned on by pressing the switch.

One important advantage of an icon is its universal expressiveness that is not constrained by the comprehension of a particular language. The wide deployment of icons in products that are intended for a global market overcomes the language barriers.

3.3.2 Data in Cyberspace

In cyberspace all data items are encoded in *bit strings*. The explanation of an Itom consists of two parts, we call them computer instructions and explanation of purpose.

Example 3.3.4: A variable in a programming language is an example for a symbol. The value of the variable corresponds to the data item at the instant of accessing the variable and the name of the variable identifies the explanation of the data. The temporal properties of this Itom are implicit—on writing a variable, the old value is overwritten and on reading a variable the last value written will be retrieved.

The *computer instructions* tell the computer system how the data bit-string is partitioned into syntactic chunks and how the syntactic chunks have to be stored, retrieved, and processed by the computer. This part of the explanation can thus be considered a machine program for a (virtual) computer. Such a machine program is also represented by a bit-string. We call the data bit-string *object data* and the instruction bit-string that explains the object data, *meta data*. A computer Itom thus contains digital object data and digital meta data. The recursion stops when the meta data is a sequence of well-defined machine instructions for the destined computer. In this case, the design of the computer and its software provides the explanation for the meaning of the data.

The second part of the explanation of data, the explanation of purpose, is directed to humans who are involved in the design, operation, and maintenance of the computer system and its software. The explanation of purpose is part of the documentation of the cyber system and must be expressed in a form that is understandable to the human user/designer.

Sometimes data is used for another than the original purpose, giving rise to many discussions about legality and data privacy of the use of the data.

If the context of an Itom changes from a machine environment to a human environment, then the representation of the data and the explanation must also change to suit the opportunities and constraints of human perception and human cognition, still keeping the semantic content of the Itom invariant. The information carried by an Itom does not require any particular representation of the data in the Itom.

3.3.3   Explanation and Generation of Data

The explanation of the data retrieves the contents—the meaning— of an Itom (the information) out of its data and the explanation provided by the shared context. In human communication, the context is given by the already existing concepts in the mind of a human receiver and in the common environment of sender and receiver. In machine communication the context is given by the rules for interpreting the data items by the software and the design of the machine. The context determines the structure and meaning of the signifiers that are used in the representation of an Itom.

The explanation of the data that is supposed to bring back the original meaning of an Itom must assume that the related prior complementary process of generating the data is truthfully reflected in the explanation.

This is an important issue when analyzing a given data set where the purpose of the data generation, the temporal attributes of the data and the precise process of data generation are unknown. There is often a certain bias in a data generation process. It is a wise habit to carefully investigate the details of the data generation process before analyzing a large data set.

The portrayal of an Itom (see Fig. 5.1), consisting of the data and the explanation of the data in the Itom, must give answers to the following five central questions:

Identification What entity is involved? The entity must be clearly identified in the space-time reference frame, established by the *Universe of Discourse (UoD)* and the Interval of Discourse (IoD). With the coming Internet of Things (IoT), the magnitude of the UoD in cyberspace is vastly expanded. (See Sect. 11.1 on the notion of an entity).

Purpose Why was the data created? The answer establishes the link between the *sense data* and the *refined data*. The question is addressed to a human user or to the designer of a computer system that processes data. The concept of purpose is alien to a machine processing information. Depending on the purpose, we distinguish between archival data and control data. Archival data reports about an observation at a past instant that is collected for archival purposes. Control data is concerned with the ongoing real-time control of a process in the environment. Control data loses its validity as real-time progresses.

Meaning What is the meaning of the data to a human or how must a machine manipulate the data? If the answer to this question is directed towards a human, then the presentation of the answer must use symbols and refer to concepts that are familiar to the human. If a computer acquires data, then the explanation must specify how the data has been acquired and must be manipulated and stored by a computer. The explanation of an actuator must describe how the actuator translates the output data into a physical action.

Time What are the temporal properties of the data? Depending on the chosen Interval of Discourse (IoD). We can classify data as static or dynamic. Archival data of a dynamic entity must include the instant of observation in the Itom. In control applications it is helpful to include a second timestamp, a validity instant that delimits the temporal validity of the control data as part of the Itom. If output data from a computer is transferred to a machine (e.g. a control valve) it must be specified at exactly what point in time the output data has to be presented to the machine.

Ownership Who owns the data? The question of ownership of personal data relates to the legal issues about privacy and is a topic of intense societal debate. Additionally, an explanation might contain evidence that helps to establish the authenticity and integrity of the data contained in an Itom (e.g., cryptographic keys). Encrypted data (called cipher text) is data that has been subjected to a transformation (encryption) such that an unauthorized user cannot easily interpret the encrypted data.

The above five questions must also guide the data generation process.

To facilitate the exchange of information among heterogeneous computer systems in the Internet, markup languages, such as the *Extensible Markup Language XML* [WWW13] that help to explain the meaning of data have been developed. Since in XML the explanation is separated from the data, the explanation can be adopted to the context of use of the data. Markup languages provide a mechanism to support the explanation of data in an Itom.

3.3.4   Sense Data and Refined Data

A Cyber-Physical System (CPS) consists of a physical object and a computer system that controls this physical object by the algorithmic manipulation of the acquired data. Sensors form the interfaces between cyberspace and the physical objects on the input side, while actuators form the output interfaces.

A sensor is a physical device that transforms a physical quantity from the physical world to sense data in cyber space. The sensor can be a human sense organ or a technical sensor. We call the physical quantity that impacts a sensor in the physical world a *physical sensation*.

Example 3.3.5: The physical sensation of a physical object that impacts a sensor does not only depend on the properties of the physical object, but also on the physical environment of the object, e.g., the illumination of the object.

A sensor thus transforms a physical sensation to sense data. In cyber-space the immediate output of a sensor, the sense data, is a *bit pattern*. The explanation of the sense data depends on the idiosyncratic design of the sensor.

Example 3.3.6: A *virtual reality system* is a CPS that includes sensors to acquire some properties of the physical environment and contains algorithms that produce a virtual world, considering the acquired sense data. The outputs of these algorithms are transformed by actuators to physical patterns that provide physical sensations for human sense organs (e.g., the eye or the ear). An *augmented reality system* enhances the physical sensations from the physical environment with physical sensations generated by a CPS.

The same physical sensation can be measured by very different physical sensing techniques, giving rise to very different bit-patterns—different sense data—as the immediate output of the sensor system. In order to simplify the further processing, the sense data should be transformed to a standardized form, called the refined data.

Example 3.3.7: Take the example of temperature measurement. The physical quantity temperature can be measured by a thermometer, a thermocouple, a temperature sensitive resistor or by an infrared sensor giving rise to very different idiosyncratic formats of the sense data. However, the refined data that is produced by the sensor system after processing the sense data is represented in a standard format, e.g., degrees Kelvin.

Every sensor can fail—in order to mitigate a sensor failure, redundant sensors must be provided in a safety-critical application. Refined data eliminates this redundancy and abstracts from those properties of the sense data that are not relevant for the given purpose (but maybe relevant for another purpose). The representation of the refined data is determined by the given purpose and the architectural style, i.e. conventions and standards in the given context (see also Sect. 9.2). The purpose defines the viewpoint (or perspective) for the data-abstraction from the sense data to the refined data. The data-abstraction determines what properties of the sense data may be disregarded in the data-refinement process for the given purpose.

Example 3.3.8: Consider the example of a camera that monitors the traffic on a highway for the purpose of toll collection. The image taken by a camera (the sense data consisting of millions of pixels) is processed to locate and identify the license plate of a car (refined data). The license plate is the signifier part of a symbol that indirectly designates the driver's bank account—the designatum—from which the highway toll can be deducted.

From the point of view of complexity reduction, the representation of the idiosyncratic sense data should be hidden within a sensor interface component of the sensor system. The sensor component includes the physical sensors and the preprocessing software of the sense data such that only the standardized form of the refined data is visible at the user interface. However, care must be taken that the given precision of the sense data, determined by the physical sensing process, is not diluted during the preprocessing of the sense-data.

Example 3.3.9: The purpose of computer aided tomography (CAT) is the detection of anomalies in the anatomy of a patient. In the first phase of a CAT scan many X-ray measurements are taken from different angles of the object, resulting in a large data structure of sensed attenuation values of the X-ray beams (the sense data). Digital geometry processing is used to generate out of these attenuation values a three-dimensional finite element model of attenuation values of the object. This three-dimensional model can be viewed from different angles to produce a two-dimensional image of the attenuation values of the object, an image of the refined data showing the internal structure of the object in different shades of gray. Based on his know-how and experience, the radiologist explains this image and determines whether an anomaly can be spotted.

In some systems, the explanation is supported by a computer program that has learned what a normal image is supposed to look like. Deviations from the norm are highlighted on the image by this computer program.

The acquisition of input data can often be compared to taking a snapshot of an entity to create an image e.g., taking a snapshot of a person. The sense data (i.e. the pattern of the image) is determined by the elementary discriminatory capabilities of the sensor system, e.g., the eye or the camera (a given number of pixels of varying intensity). The characteristics of the image depend not only on the properties of the observed entity, but also on the purpose that determines the position and view of the camera and the instant of taking the snapshot. The purpose has thus an effect on the characteristics of the image i.e., the acquired input data.

After (in the temporal sense) sense data has been acquired by a sensor system it can be stored on a physical entity called the *data carrier* in order that it can be accessed at a later time. The data carrier can be biological (e.g., human memory) or technological (e.g., computer memory). Imprinting data on and retrieving data from the data carrier as well as the processing of data require energy.

3.4   Essential Data (e-data) Versus Context Data (c-data)

If a speaker and the listener have many mental experiences in common then the shared details of the context (the shared context) is large and only a small set of particular details (data) has to be exchanged in order to realize a communication act (for the transmission of an Itom).

> **Example 3.4.1:** A group of friends, sitting in a pub, shout numbers at each other. Some numbers cause the group to start laughing, while other numbers are considered boring. A newcomer watches the group for a while and then asks "What is the meaning of a number?" He is being told that every number refers to one of a set of jokes that all friends know. They only laugh at good jokes, not at boring jokes.

The shared details of a context consist of common internal elements in the conceptual landscapes (the minds) of the speaker and the listener and of external elements that are provided by the external physical environment in the current situation.

> *We call the data that remains after the shared context of the sender and receiver are fully aligned the essential data or **e-data** of a communication act. Data that is required to establish the alignment of the shared context is called context data or **c-data**.*

According to Minsky [Min74, p.1] the internal elements of the shared context are stored in frames, i.e. tree-like data structure, in the mind of an individual: "We can think of a frame as a network of nodes and relations. The top levels of a frame are fixed and are always true about the supposed situation. The lower levels have many terminals—slots—that must be filled by specific instances of data." A frame thus provides the explanation for the data. When talking about language, the most important part of the internal elements is a shared view of the denotation of the words of the chosen language.

The external elements comprise location, time and other properties of the external environment, e.g., rain, sunshine etc. To arrive at an agreement about the current situation the data of the external elements must be filled in the slots of a frame.

In the case where the shared contexts among senders and receivers are aligned up to a single bit (the remaining e-data), the transmission of this single bit of data is sufficient to transmit the Itom.

Example 3.4.2:   In ancient times the information that an "enemy is approaching" was transmitted by the incineration of fires on a hill, thus encoding the vital single bit of data— the e-data— in the appearance of a pre-arranged fire. The c-data, i.e., the context for the interpretation of the fire has been established ahead of time and does not change.

The smoke signal (the signifier of an Itom) that appears at the chimney of the Sistine Chapel during the election of a new pope is more involved. The rules for the interpretation of the smoke (the c-data) are well known. The occurrence of the smoke signal informs that an election has been taken place. The color of the smoke (white or black) informs whether the election of a new pope has been successful.

The responsiveness of a distributed control system can be improved and the load on the communication system can be reduced, if the bit-pattern (the data) that must be transported from one component to another component of the distributed control system is minimized.

The distinction between e-data and c-data becomes important when the size of a real-time message must be reduced in order to minimize the transport latency.

A separation of the e-data and the c-data within an Itom can help to shorten the bit pattern (the data) that must be transported in a communication act in order to transfer an Itom. Ideally, only the e-data of the Itom must be transported and the c-data can be acquired locally from the context.

Example 3.4.3:   Many years ago—before the time of real-time computer networks—the problem of the transmission of sensor data (the e-data) to the control room was solved by the implementation of a 4–20 mA current loop between each sensor and the control room, where 20 mA meant 100% of the specified range, 4 mA meant 0% of the specified range and 0 mA meant line broken. The identification of the Itom—part of the c data—was implied from the context, i.e., the wiring. No explicit time stamping was necessary, because the *transport latency* of a current loop can be neglected.

When real-time computer networks are installed to transmit sensor data, then, in addition to the value of the sensed variable—the e-data—, the contextual information—the c-data—, i.e. name and location of the entity that is observed and the instant of observation must be provided to enable the receiver to arrive at the meaning of the data.

In a time-triggered system the scope of the shared context of the sender and the receiver is enlarged by the provision of a global time. This global time can be used to reduce the amount of c-data that must be transmitted between a sender and the receivers.

In time-triggered system, static c-data can be provided locally at the receiver by linking it to the progression of the shared global time.

Example 3.4.4:   In a time-triggered system, the explanation of the e-data (contained in a periodically arriving message) can be linked locally at the side of the receiver to the *a priori* known periodic instant when the message arrives.

The global time has been used in the design of the time-triggered protocol TTP to reduce the transmission of c-data and to provide a fault-tolerant clock synchronization without any extra load on the communication system (see Sect. 8.3).

The amount of data in an Itom and the alignment of the shared context are in an inverse relation—we call this the *reciprocity principle*. We can thus reduce the amount of data that needs to be transferred in a communication act by improving the alignment of the shared context. There is a trade-off between the efforts required to build-up a refined shared context and the gain in the efficiency of communication.

In the previous analysis we assumed that the view of the shared context is exactly the same at the speaker and listener. In general, this assumption is not fully justified. We therefore must distinguish between the view of the shared context by the sender and the view of the shared context by the receiver. Differences between the view of the shared context by the sender and the view of the shared context by the receiver are the reason for many misunderstandings. These differences are occasionally caused by differences about the denotation of a word—which are relatively small within a language community—but more often by the differences in the connotation of a word—which depends on the different personal and emotional experiences associated with a word by the sender or by the receiver.

3.5   Stigmergy

The definition of communication, cited at the beginning of Sect. 3.2 (taken from the Merriam-Webster dictionary) contains in the last sentence the comment: *"The function of pheromones in insect communication."* This comment refers to the method of *stigmergic information transfer.*

> *We call the transmission of information between a sender and a receiver via observable modifications of the state of a shared physical environment a stigmergic information transfer.*

The biologist Grasse [Gra59] introduced the term *stigmergy* to describe the indirect information flow among the members of an ant colony when they coordinate their activities.

Example 3.5.1: Whenever an ant builds or follows a trail, it deposits a greater or lesser amount of the whiffing chemical pheromone on the trail, depending on whether it has successfully found a prey or not. Due to positive feedback, successful trails—i.e. trails that lead to an abundant supply of prey—end up with a high concentration of pheromone. The nearly blind ants capture by their olfactory sense the intensity of the pheromone smell. It has been observed that the running speed of the ants on a trail is a non-linear function of the trail-pheromone concentration. Since the trail-pheromone evaporates — we call this process environmental dynamics — unused trails disappear autonomously as time progresses.

A stigmergic information transfer between one or more senders and one or more receivers is realized in the following steps:

(i) A sender sends a message to an actuator that alters the state of the physical environment.
(ii) This altered state of the physical environment is possibly modified by *environmental dynamics* resulting in a *modified altered state*.
(iii) A sensor of the receiver observes this modified altered state by a sensor system and produces sense data.
(iv) The sense data is combined with the information retrieved from the context to generate one or more Itoms.

There are three fundamental differences between information transfer by messages and a stigmergic information transfer.

The first difference relates to the requirements on the shared context between sender and receiver. As noted before, the shared context consists of external elements that are provided by the external physical environment and of common internal elements in the conceptual landscapes (the minds) of the sender and the receiver. The external elements, such as location, time and other properties of the environment, form the major part of the shared context in a stigmergic information transfer. There is only a limited need for common internal elements among the sender and the receiver in a stigmergic information transfer.

> **Example 3.5.2:** A good example for stigmergic information transfer is the information transfer among the drivers of cars at a busy intersection. Although the drivers can come from different backgrounds and have limited shared internal elements in their conceptual landscapes (e.g., they don't even speak the same language) they can communicate by observing the changes in the state of the environment caused by the actions of other drivers on the road.

The second difference relates to effects of *environmental dynamics*. In a stigmergic information transfer the effects of environmental dynamics on the state of the environment are part of the communication process. In contrast, message-based information transfer occurs in cyberspace and does not consider environmental dynamics.

> **Example 3.5.3:** If two fireman exchange electronic messages about the state of a burning building, they do not grasp the current state of the fire that is conveyed by a look at the fire (stigmergic information).

The third difference relates to the roles that the actuator system and the sensor system play in a stigmergic information transfer. The design of the actuator system determines which property of the environment is altered. The design of the sensor system determines the property of the environment that is sensed and contained in the sense data of the receiver.

If, in a Cyber-Physical System project, the focus of the system analysis is restricted to the cyberspace, then the stigmergic information transfer realized by the physical environment is overlooked.

.

Invisible and unattended information flows between identified subsystems pose a considerable barrier to understanding. Case studies about the understanding of the behavior of large systems have shown that the perceptually available information plays an important role for developing an understanding of a system [Hme04]. Unattended information flows can also play an important role in the creation of emergent behavior (see Sect. 5.4).

3.6 Points to Remember

- A symbol is a sign where the relationship between the signifier (the pattern of the symbol) and the designatum (the meaning of the symbol) is determined by the context, i.e., cultural circumstances and conventions that must be learned. A word is an example for a symbol.
- An explanation of a symbol establishes the link between the signifier of the symbol and the designatum. The signifier of a symbol is meaningless if there is no explanation of the symbol available.
- An Itom is a semantic communication construct that informs about a proposition of a state or a sequence of states of an entity in the world at a specified instant in time. It is represented by an atomic triple consisting of a data item (the signifier part of a symbol), the explanation of the data item and the time that denotes when the proposition holds. An Itom is created for a purpose by a human author.
- A data structure consists of a collection of related data items.
- The representation of a data item is determined by the given context, while the semantic content carried by an Itom has a connotation of absoluteness. The information carried by an Itom does not depend on any particular representation of the data in the Itom.
- In human communication, the explanation of a data item (e.g., the meaning of a word) is derived from the shared context between sender and receiver. In machine communication, the explanation of a data item is given by the rules for processing the data by the machine.
- The explanation of the data that brings back the original meaning of an Itom must assume that the related prior complementary process of generating the data is truthfully reflected in the explanation.
- If a speaker and a listener have many mental experiences in common then the shared details of the context (the shared context) is large and only a small set of particular details (data) have to be exchanged to realize the transmission of an Itom.
- We distinguish between c-data (context data) that is used to build up the shared context and e-data (essential data) that must be communicated if the shared contexts between speaker and listener are fully aligned. The c-data is normally static, while the e-data is dynamic.

- The explanation of an Itom destined for a computer tells the computer system how the data bit-string is partitioned into syntactic chunks and how the syntactic chunks have to be stored, retrieved, and processed by the computer.
- A sensor is a physical device that transforms a physical quantity in the physical world to a bit-pattern, the sense data, in cyberspace. Refined data eliminates redundancy and abstracts from those properties of the sense data that are not relevant for the given purpose (but maybe relevant for another purpose).
- A stigmergic information transfer refers to the transmission of information between a sender and a receiver via observable modifications of the state of a shared physical environment.

Chapter 4
Modeling

> *"Modeling is the art of simplification and abstraction, taking only "so much" from reality to answer the question put forth at the right abstraction level."*
>
> —Saurabh Mittal, Saikou Diallo, Andreas Tolk

4.1 Introduction

Any system in the real world—a *natural system*—is characterized by an immense number of different properties. Some of these properties, e.g., radioactivity, cannot even be perceived with human sense organs. The cognitive apparatus of humans cannot handle the totality of all these properties. This is the reason why humans construct models of natural systems that focus on only those properties that are relevant for a given purpose and forget about the rest.

A model of a system is a simplified representation of the system that is built in order to understand a given system for a given purpose. We call the system that is being modeled the target.

The target can be a set of observations that needs explanation, a thing, a process or some other model. The purpose of the model determines which properties of the target must be part of the model and which properties can be ignored. It is impossible to build a model of a target without a crystal-clear statement about the purpose of the model.

Depending on purpose, there can be widely different models of the same target.

Example 4.1.1: Consider the different models of a computer implemented on a VLSI chip:

- A model of the instruction set implemented on the chip in order to examine the execution of algorithms on this computer.
- A model of the timing of the interconnect on the chip in order to verify if the physical layout meets all timing requirements.

© Springer Nature Switzerland AG 2019
H. Kopetz, *Simplicity is Complex*, https://doi.org/10.1007/978-3-030-20411-2_4

- A model of the distribution of dopants in the semiconductor of the chip in order to improve the switching speed of the transistors.
- A model of the temperature profile inside the chip in order to locate hot spots.
- A model of the physical dimensions of the chip in order to study different forms of packaging.

Models of reality are always idealizations, i.e. deliberate abstractions, in order to make reality tractable. Example 4.1.1 demonstrates that a particular model of a thing is concerned with only one or a few properties of the thing. By restricting a model to a few properties, we achieve the required simplification. "Models are always incomplete, they are simpler than the entities they represent [Joh83, p.10]."

In our conceptual landscape we have a multitude of useful *mental models* of relevant aspects of reality. Whenever different models can explain a set of observations the model that relies on the smallest set of assumptions—the *simplest* model— wins and determines the accepted view. This *principle of parsimony* of William Ockham plays an important role in the history of science [Sob15].

Example 4.1.2:   A given phenomenon, e.g., the set of observations of the movements of the stars on the sky, may be explained using different models. For a long time, the prevalent model for the explanation of the movement of the stars was the geo-centric model of Clausius Ptolemy, where it is assumed that the earth is at the center of the universe and all planets move in epicycles around the earth. In 1543 Nicholas Copernicus proposed a helio-centric model in which the earth and the other planets move around the sun. In [Sob15, p.12] it is stated: *"There was no observation that Copernicus could cite that refuted the Ptolemaic model."* For Copernicus, the only reason for preference of his helio-centric model was that the helio-centric model is simpler than the geo-centric Ptolemaic model, following William of Ockham's important principle of parsimony (referred to as *Ockham's razor*). What is the true nature of reality, if two vastly different models, both with identical explanatory power, explain a set of observations?

The decision, which subset of the many properties of a target is selected to become part of a model depends on the given purpose of the model.

From the point of view of a Cyber-Physical System, a model of behavior of the target that has predictive and explanatory power is of particular relevance.

Models of behavior are based on observations of the target, i.e. reality or another model, during an interval of time. "In order to make an observation one must have an idea of what could be seen, and a framework of beliefs in which new observations, both confirming and disconfirming, may be interwoven [Ahl96, p.13]". Modeling is thus an *iterative bootstrapping process*, starting from some idea, a simple model that forms the basis for further observations of the target to refine or refute the simple model in the next iteration.

There are two basic techniques available for constructing a model of behavior of a target, the input-output technique and the state variable technique.

Input-output technique   The input-output technique considers the target as a black box that reacts to temporal sequences of inputs with temporal sequences of outputs that can be observed. From a large number of observable input-output relations of the target, a model of the target is constructed such that the behavior of the model mimics the observed behavior of the target. If a model design is based on this input-output technique, then those properties of reality that have been observed during model construction are included in the model.

Whenever a situation occurs that has not been observed during the learning phase, then the predictions of a black-box model are not trustworthy.

This important insight must be kept in mind when machine learning, e.g. by deep neural networks, is used to arrive at a model [Liu17].

Example 4.1.3:   In the domain of medicine controlled clinical studies are conducted to establish the reaction of the organism as a consequence of the intake of an identified medication (Input-output technique). If a group of persons (e.g. pregnant women) is not adequately represented in the clinical study, nothing can be said about the occurrence of side effects in this group of persons.

State-variable technique   In the *state-variable technique* (also called white box design) it is assumed that known general laws according to the HO schema determine the future behavior of the target (see Sect. 2.3). At a selected instant the state of a model of the target can be described by the set of values that are assigned to the relevant properties of the target. We call a variable that denotes a selected property of a target a state variable and the set of values that are assigned to all state variables at the selected instant the state of the model at that instant. The state of a model at a chosen instant thus incorporates all inputs of the past (i.e. before this instant) that can have an effect on future behavior (i.e. after this instant). In an ideal deterministic model, the state of the target at a selected instant denotes the initial and boundary conditions that, together with the general laws, determine all future states of the model of the target, i.e. the *deterministic trajectory* of the evolution of the target in the temporal domain. Since the general laws that are at the core of the state-variable modeling technique are universally valid and are considered true at all points in time, this modeling technique supports also predictions about the future behavior of the target in novel circumstances that have not been observed in the past.

Example 4.1.4:   The trajectory of a rocket to the moon can be modeled by the state-variable technique.

In some applications where only a partial knowledge of the internal working of a system is available, a combination of the state-variable technique and the input-output technique is performed in order to arrive at a hybrid or gray-box model [Kha12]. This combination of theoretical knowledge with the results of the analysis of empirical data is a most promising approach to model building.

It is the purpose of this Chapter to discuss the relation between a target and the models of behavior of the target and to elaborate on the differences between a rational model, an intuitive model, a scientific model, a pragmatic mental model and a cyber model of behavior.

4.2   Modeling Relation

We call the relation between a target and the model of the target the *modeling relation* [Ros12, p.85]. The modeling relation links the target (a thing in the physical world or another model) to the model world. A target in the physical world must obey the natural laws, e.g., the laws of physics. A model resides in a symbol world (in the mind of a person or in cyberspace), follows the rules of this symbol world and imitates the behavior of the target by *symbol manipulation*.

Let us assume that the symbol world is characterized by a static and uniform context, i.e., the relations between a signifier (the presentation of a symbol) and the associated designatum (the meaning of the symbol) are known and constant. It follows that a manipulation of the signifier implies a corresponding manipulation of the designatum of the symbol. This is symbol manipulation.

> **Example 4.2.1:**   Assume that the character a (the signifier of a symbol) stands for the length of the side of a square (the designatum). The area of the square is then given by $a.a = a^2$. By manipulating the signifier, we arrive at a property of the designatum (the area of a square). This operation is only valid, if the signifier refers to a square.

Starting from corresponding initial states, the target and the model work in some related way during a defined interval of real time and produce—in the ideal case— at the end of this given time interval the same resulting states. Technical or biological sensors and actuators form the interfaces between the physical world and the symbol world.

> A *sensor encodes a property of a system in the physical world (a physical quantity) into a symbol of the symbol world, while an actuator transforms a symbol of the symbol world into a physical action in the physical world.*

We call the concrete steps of symbol manipulation, prescribed by the model, that lead from a start state to an end state in the symbol world the model algorithm. An algorithm per se is timeless, although the execution of the algorithm, e.g. on a digital computer, takes a certain real time-interval (the execution time of the algorithm), the length of which depends on the characteristics of the algorithm and the performance of the computer.

Let us now investigate the progress of a physical process (the target) and the progress of the model of the target over a defined physical time interval—we call it a *time-slice* (Fig. 4.1). At the start of a time-slice a sensor system observes all properties of the *target* that are relevant for the given purpose and records this start-state in state variables of the model in the symbol world (Arrow 1 in Fig. 4.1). During the time slice two concurrent activities take place: (i) the target progresses from the start of the time slice to the end of the time-slice according to the applicable laws, e.g., the laws of physics, and produces *a physical end-state* (Arrow 2 in Fig. 2.1) and (ii): the *model process* that contains the explanation of the data in the state variables and the rules for the evolution of the target takes the symbols of the start-state and the duration of the time-slice as inputs and calculates by the execution of the model

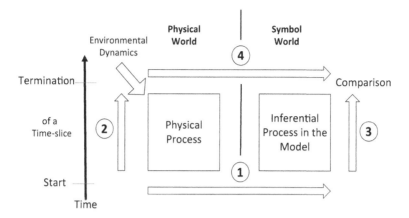

Fig. 4.1 Modeling relation

algorithm (Arrow 3 in Fig. 4.1) the model end-state in the symbol world. At the end of the time-slice the sensor system observes all relevant properties of the target again (Arrow 4 in Fig. 4.1) and produces the symbols of the physical end state in the symbol world. If the model is perfect then the physical end-state and the model end-state in the symbol world will be alike. Note that the physical duration of the time slice must be known *a priori* of the model execution, because this duration is an important input to the model algorithm that determines the point in time for which the predictions of the model must be made.

Let us now assume that the arrow of real-time is partitioned into a sequence of equidistant time-slices, such that the end of one time-slice coincides with start of the following time-slice. In the parlance of control theory, a start-point of a time-slice is called a *sampling point*, the duration of the time-slice is called the *sampling interval* and a time series of state variables is called a *trajectory*. In *receding horizon control* the above-described procedure is executed in each time-slice of the sequence, such that the prediction horizon is shifted forward towards the future. If we observe the target at every sampling point, we generate a periodic data stream of observations and model outputs, i.e., trajectories of the observed and calculated variables.

The communication between the sensors that observe the physical world and the computer system that executes the model algorithm is best realized by a *periodic time-triggered protocol* that provides a guaranteed minimal transport latency (see Sect. 8.3).

A model is called *deterministic* (see also the definition of *determinism* in Sect. 2.3) if a given input, consisting of the input symbols and the duration of the time-slice will always produce the same output symbols. Determinism of a model is a useful property of a model, since it helps to reason about the behavior of the model and thus supports the certification of a model.

The duration of the time-slice synchronizes the progress in the model world with the progress of the target in the physical world. If the execution time of the model algorithm is much smaller than the duration of a time slice, predictions about the behavior of a deterministic target in the distant future can be made.

4.2.1 Execution Time of Model Algorithms

The amount of time that is provided for the execution of the control algorithm on the given hardware platform is called the *service interval (SI)*. The duration of the SI depends on the number of steps of the algorithm and the performance of the computer that executes the algorithm. The duration of the time slice sets an upper limit on the duration of the SI.

Many algorithms are required to run to completion before they deliver a useful result. We call these algorithms *WCET* (*worst case execution time*) algorithms. In the WCET approach algorithms are developed where the worst-case execution time (WCET) bound for the algorithm execution on the given hardware base can be established *a priori* (at design time) on the given hardware platform for all data points of the input domain [Wil08]. This produces a conservative design because the fight for a tight bound on the WCET has two enemies: an enemy from below and an enemy from above. The enemy from below refers to temporal indeterminism that is inherent in modern hardware architectures. The enemy from above refers to algorithmic issues, e.g., the complexity of a computationally expensive algorithm that makes it difficult or even impossible to establish a WCET bound for all data points of the given input domain.

In the domain of real-time control another type of algorithm, we call it an *anytime algorithm,* is more appropriate [Kop18]. An anytime algorithm consists of a *core segment* followed by an *enhancing segment*. The execution of the core segment is guaranteed to provide a first satisficing result quickly. A result is satisficing if it is good enough in the current scenario—not necessarily optimal—and meets all safety criteria. Continuous improvements of the satisficing result are provided by the repeated execution of the enhancing segment until the end of the given service interval (SI) on the available hardware is reached.

Example 4.2.2: A good example for an anytime algorithm is Newton's *iteration algorithm.*

An anytime algorithm thus makes always best use of the available execution time slot, the service interval (SI). In the anytime approach the WCET of the core segment must be smaller than the duration of the service interval (SI). As a consequence, the core segment must deploy algorithms that are amenable to WCET analysis. The substantial time-interval between the average execution time of the core segment and the WCET of the core segment is used to improve the quality of the result. The WCET bound for the core segment can be derived either from an analytical analysis of the core-segment code of the algorithm [Wil08] or from experimental observations of the execution times of the algorithm in the given application context (or from both). In order that the model process does not fall behind the target, we must make sure that service interval (SI) on the available computer hardware is smaller than the duration of a time-slice., i.e., the sampling interval.

4.2.2 Model Error

There exists no perfect model of a physical thing for the following reasons:

- It is impossible to include all properties of a thing in a single model.
- It is impossible to consider all environmental effects in the real world that can have an impact on the operation of a physical thing—the target. We call these diverse environmental effects that are not explicitly considered in the model *environmental dynamics*. Consider, e.g., the earthquake cited in example 2.3.4 during a game of billiards!
- The model algorithm may be based on insufficient knowledge about the laws that control the physical process. For example, a non-linearity is not properly modeled.
- The service interval SI provided for the execution of an anytime algorithm is too short to find a good solution.

We call the difference between the physical end-state and the model end-state at the end of a time-slice the *model error*. The model error can be reduced by

- Shortening the duration of the time-slice and thus the prediction horizon of the model.
- The extension of the model by the addition of sensors that measure the impact of environmental dynamics.
- The refinement of the model algorithm.
- An improved performance of the computer platform.

A shortening of the time-slice implies that the time-interval for the execution of the model algorithm is also reduced. Depending on the dynamics of the physical process and the given available performance of the machine that executes the model algorithm, there exists an optimal duration for a time slice in a given modeling environment [Kop18]. An increase in the performance of the computer shortens this optimal duration of the time-slice and thus reduces the model error.

An essential difference between science and engineering can be clarified by looking at the modeling relation of Fig. 4.1 [Lee17]. A scientist observes the properties of a natural system and constructs a model of this natural system that mimics the behavior of the natural system in order to gain a profound understanding of the observed phenomena. A scientist thus proceeds from the left to the right of Fig. 4.1. An engineer proceeds in the other direction, from right to left. He first creates a model of an innovative artifact that constitutes the plan for the subsequent realization of this artifact.

Ed Lee makes an important remark about the relation of science to engineering in [Lee17, p.18]: "As any engineer will tell you, innovations, such as aviation or the

steam engine, commonly precede scientific understanding of how things work."
This is in contrast to the prevailing opinion that scientific understanding precedes
innovation.

4.3   Mental Models

Humans have a multitude of models of their own behavior and of the perceived
structure and behavior of their environment—the external reality— in their minds in
order to orient themselves in the world. We call these models *mental models*. In the
book *The Nature of Explanation* Kenneth Craik posits [Cra67, p.61]: *"If the organism carries a 'small scale model' of external reality and of its own possible actions
within its head, it is able to try out various alternatives, conclude which is best of
them, react to future situations before they arise, utilize the knowledge of past events
in dealing with the present and future, and in every way to react in a much fuller,
safer and more competent manner to the emergencies which face it."* According to
Craik [Cra67, p.50] mental reasoning about a physical phenomenon—i.e., a process
in the external world—with the aid of a mental model can be broken down into three
steps:

- *"'Translation' of external process into words, numbers or other symbols,*
- *Arrival at other symbols by a process of 'reasoning' deduction, inference etc.,
  and*
- *'Retranslation' of the symbols into external processes (as in building a bridge
  to design) or at least recognition of the correspondence between these symbols and external events.*

*One other point is clear: this process of reasoning has produced a final result
similar to that which might have been reached by causing the actual physical process to occur."*

Step (2), of the above schema, the *Arrival at other symbols by a process of 'reasoning' deduction, inference* refers to the execution of a *mental model* of the physical phenomenon which requires *cognitive effort*.

The modeling relation of Fig. 4.1 shows the first and second of these steps of
reasoning, (Arrow 1 and Arrow 3 of Fig. 4.1). If we retranslate the symbols of the
symbol world into effects in the *external world*, we arrive at, what Johnson-Laird
[Joh83; p.403] calls a *Craikian Automaton* (Fig. 4.2).

Arrow 1 in the *Craikian Automaton* (Fig. 4.2) denotes the biological sensors that
transform the representation of the *Itoms* of the physical world to the representation
of the *Itoms* (see Sect. 3.1) in the symbol world of the human mind and Arrow 4
denotes the biological effectors that transform the representation of the *Itoms* in the
symbol world to actions in the physical world. The box *Representation of the World*
contains the conceptual landscape or the knowledge *base* that has been built up in
the human mind, part of it *innate* and parts of it by *lifelong learning and experience*
(see also Sect. 2.2). Arrow *3* refers to the execution of the mental model, retrieved

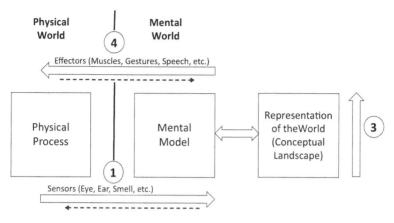

Fig. 4.2 Craikian automaton

from the knowledge base that manipulates the acquired Itoms and produces the intended Itoms for the output to the effectors. The execution of the mental model is a process of thought, an *inference* that leads from the set of input Itoms (propositions) to the output Itoms.

The dotted lines in arrows *(1)* and *(4)* picture low-level feedback mechanisms that help to control and enhance the *human perception* realized by the human senses and *human actions*. These low-level feedback mechanisms are important for the improvement of the quality of a *sense action* and an *actuation* action. During a *sense action*, the intentions of the human observer, his assumptions about the current context and his initial hypothesis about the sensed scenario are brought into the sensor to control, refine and retarget the sense actions (e.g., the *saccades* of the eyes). During an *actuator action*, the reactions of the environment to the physical forces of the actuator are perceived in order to verify the assumptions about the current scenario and to control the applied forces.

Every mental model in the conceptual landscape of a person can be placed on a *continuum* between two extremes, a purely *rational model* on one side that is executed in the rational subsystem of the mind and on the other side a purely *intuitive model* that is executed *unconsciously* in the intuitive subsystem (see also Sect. 2.5). In-between these two extremes are the *pragmatic mental models* (PMM) that are partially *rational* and partially *intuitive*.

4.3.1 Rational Model

Rational models are part of the *rational system* of the human mind. The execution of a rational model requires *explicit cognitive control* that must often be expended over a prolonged time interval. A cognitive effort—similar to a physical effort—can evoke negative emotions and fatigue, such that a person tries to avoid this effort.

Example 4.3.1:   The *logical interferences* that take place in the *analysis of a mathematical proof* are a good example for a rational model.

Scientific models are a special class of rational models. In scientific models the scope of the model is precisely defined—normally an *ideal setting* (see Sect. 2.2) is assumed—, all assumptions are explicitly listed and any element of subjective input is avoided.

Example 4.3.2:   A formal theory that generalizes concrete experiences, explains diverse phenomena in a domain, and provides the capability for predictions in novel situations is an example for a *scientific model*.

The acquisition of an abstract theory requires extensive training that is associated with a significant *cognitive effort*. On the other side, the familiarity with such a theory can guide human behavior in unchartered territory and help to make a quick decision in an unusual scenario. The cognitive effort spent in the learning of such a formal theory is paid back by the reduction of the *cognitive effort*—and thus the reduced cognitive complexity—in situations where the theory can be applied.

In Sect. 2.5 we have discussed a number of biological limitations of the cognitive apparatus for analyzing rational tasks. These limitations concern constraints of storage and processing in short time memory and fatigue, caused by metabolic mechanisms. A rational mental model for the solution of a problem can only be handled by the rational system of the human mind if its representation fits into the finite size of the cognitive apparatus of a human. If the model is too large or too complex, we have to reduce its size and complexity by the application of simplification strategies. Some of these simplification strategies are discussed in the second part of this book.

4.3.2   Intuitive Model

In an *intuitive model* implicit inferences that lead to the desired outcome are executed quickly in the intuitive subsystem of the human mind without explicit cognitive control. The human response time can be as low as 100 msec. For example, an intuitive model controls walking, and the capability to walk is acquired without precise instructions. The same holds for effortless implicit inferences that are the basis for intuitive judgments and language comprehension. The analytical study of the sensory and control actions that take place during the execution of an intuitive model is very difficult and often impossible to perform—take, for example, the control of the muscles in the articulation of speech. Intuitive models are part of the intuitive subsystem of the human mind (see Sect. 2.2).

Example 4.3.3:   It is alleged that the swift detection of a snake by a human is realized by special intuitive *snake detection neurons* without the involvement of the rational subsystem of the human mind [Ben13].

There are many scenarios, such as learning a new sport or learning a new language, that require explicit cognitive control and a significant cognitive effort during the learning phase and during the initial executions, but where over time repetitive executions, drill and practice move the model from the rational side to the intuitive side. When the proficiency of the person increases the amount of cognitive resources required for the execution of the model is reduced and the *rational cognitive apparatus* is freed for other tasks.

An expert in a field has mastered this transition of the relevant domain knowledge from the rational subsystem to the intuitive subsystem. It takes a long time—more than 10 years—of hard work to become an expert in any domain, be it science, music, or any sport [Eri06]. Whereas a layman may spend a long time finding a solution to a given problem, a domain expert can solve this problem intuitively and instantly—this is our measure of *cognitive simplicity*. The hard cognitive work over an extended period of time is thus rewarded by simplicity of understanding of novel phenomena in the domain. (See also example 2.5.6 about chess playing).

4.3.3 Pragmatic Mental Model (PMM)

In order to make best use of the limited and costly cognitive capacity, the mind carefully controls the allocation of cognitive resources to competing tasks based on a *cost-reward analysis* [She17]. The proper allocation of the limited cognitive resources is essential to the survival in a complex world. Short-term rewards and long-term rewards (or punishments) are competing for the limited cognitive resources. In order to reduce the cognitive effort and provide a solution to a problem within given critical time constraints, humans are often forced to resort to a *pragmatic mental model (PMM)* that leads quickly to a satisficing result instead of spending an extensive amount of time and effort to find an optimal result by using a larger more involved rational model. The concept of *bounded rationality of humans* is based on this observation [New72, p. 55].

Most mental models that are used to arrive at a decision in a time-constrained real-life scenario are pragmatic mental models (PMM). They are based on the principle of *bounded rationality* for the following reasons:

- It is not possible to formalize all aspects of the external environment, the environmental dynamics, and the potential behaviors of adversaries.
- The size of an all-inclusive rational mental model that considers each and every possible input exceeds the biological constraints of the cognitive apparatus of humans.
- In many situations, the time required to execute an all-inclusive rational mental model to find an optimal decision is not available.

As a consequence, the PMMs for decision making in a real-life scenario are composed of a combination of recognized causal dependencies among the perceived events (using the rational system) and an intuitive, sometimes emotional, judgment derived from past experiences (retrieved from the intuitive system), taking the immediate input from all senses (e.g., the perceived facial expressions of a partner) into account. These kinds of PMMs are non-deterministic, since the decision maker is not aware of all inputs and intuitive inferential processes that led to the reached result.

4.4 Cyber Models

A model that is designed for a digital computer to predict or control a process in the physical environment is called a *cyber model*.

Since a human designer is required to design a cyber model, there is no cyber model without a prior mental model in the mind of the designer.

Although the modeling relation of Fig. 4.1 pertains to cyber models as well as to mental models, cyber models differ from pragmatic mental models in a number of significant aspects (Fig. 4.3). Cyber models have more characteristics in common with scientific models than with PMMs.

The most important distinction between a PMM and a cyber model involves the differences in scope. It is impossible to identify all inputs that are considered in a pragmatic mental model, since a person cannot be fully aware of all the inputs that originate from the intuitive subsystem. In contrast, the cyber model is a formal rational model, where all inputs are precisely specified and limited by the available sensors. The algorithm of a cyber model most often implements a *deterministic effective procedure* that provides a solution to the stated problem in a finite number of steps. Since the technological limits that constrain the size and the execution time of a cyber model are extended every year, whereas the biological limits of the human mind remain constant, the capability of cyber models is on a steady increase compared with that of a pragmatic mental model.

Traditionally cyber models have been algorithmic models that use the state-variable technique (see Sect. 1) to find an analytical path to the solutions of the stated problem. They start from the given initial and boundary conditions and apply general laws to make a prediction.

Pragmatic Mental Model	**Cyber Model**
Undefined scope	Well-defined scope
Bounded rationality	Effective Procedure
Non deterministic	Deterministic
Size and execution time constrained by biological limits	Size and execution time constrained by technological limits

Fig. 4.3 Pragmatic mental models versus cyber models

With the advent of more powerful computers the size of the cyber models has grown to the point where a model must be decomposed into a number of subsystems that interact in numerous ways. An accurate large cyber model of a complicated physical phenomenon (e.g., the prediction of the weather) consists of many subsystems with numerous interdependencies between these subsystems. Computer simulations are performed to arrive at the prediction of these models. The results of these computer simulations can demonstrate an unexpected *bizarre* behavior that nevertheless is in perfect agreement with reality. This bizarre behavior does not come close to the behavior produced by any approximate PMM of the same phenomenon. From the perspective of a human observer, this bizarre behavior cannot be explained on the basis of the prevailing conceptualization in the PMM because the observer does not have access to a manageable mental model that provides an explanation for the bizarre phenomenon.

Example 4.4.1:   Although the average temperature around the globe is on the rise, the effects of global warming, according to climate models, cause some areas on the globe to get a harsher winter. These model predictions, which agree with the observations, are counter-intuitive [Gib18].

Although an extensive simulation of a large cyber model may *provide accurate predictions,* it lacks explanatory power and does not lead to a profound human understanding of a puzzling phenomenon. In order to achieve human understanding a new conceptualization must be developed that aggregates the essential elements that lead to this bizarre behavior into a single unit of thought. This new conceptualization abstracts from irrelevant detail and simplifies the description of the phenomenon such that a PMM of a manageable size can be constructed. New conceptualizations of this kind are at the core of scientific progress.

Example 4.4.2:   The introduction of an isolated mechanical world —an ideal setting— free from side effects and solely governed by the unique relations between the concepts mass, acceleration, power, energy and time formed the foundation of the mechanistic view of the physical world some hundreds of years ago.

In the very recent past significant advances in the field of *deep neural networks* have led to AI (artificial intelligence) based cyber models for pattern recognition that can be trained on large data sets and achieve a performance that comes close to or exceeds human performance in some pattern recognition tasks. These models have characteristics that are similar to large simulation models—they predict, but do not explain to humans since there exists no PMM in the rational subsystem of the human mind that can understand what is happening in these models. These deep neural networks models, based on the input-output technique, are closely related to the functioning of the intuitive subsystem of the human mind that is outside our rational comprehension of the world.

4.5   Models of Time

The model of time in a digital computer is *discrete*—time progresses in steps (e.g., *time slices*) delimited by the ticks of digital clock. During the *duration* of a step (a *granule of time*) the progression of time in the model is frozen and at the end of a step time changes abruptly by one basic unit. In contrast to this, the *Newtonian model of physics* is based on an analog model of time. A target can exhibit *continuous behavior* (e.g., the distance travelled by a free billiard ball is a function of the continuous progression of real-time) or *instantaneous behavior* (e.g. the collision of two billiard balls occurs at an instant of time).

> *Neither continuous nor instantaneous behavior can be precisely expressed by a discrete model of time.*

We have to live with—hopefully good—approximations in the digital world.

A model manipulates symbols that imitate the behavior of a target in an interval of real-time, the *physical time slice* introduced in Sect. 4.2. Since there is no explicit notion of the *progression of real-time* during a time slice in a symbol world, some implicit mechanisms must be provided to bring the temporal dimension of behavior in the physical world to the manipulation of symbols realized by the algorithm of the digital computer in the symbol world.

In Fig. 4.1—the modeling relation—we introduced the notion of a *time-slice,* a *granule of time*, where at the end of the time slice the *physical behavior of the target* and the *imitated behavior of the target* calculated by the model are compared. The *duration of the time-slice* is thus the *symbol* that brings the progression of the physical real-time into the symbolic model world. When designing a model, the selection of a proper duration of this time-slice is of paramount importance. Let us explain this on the example of the already introduced game of billiards.

Example 4.5.1:   The maximum speed of the *cue ball* immediately after the strike by the cue stick is about 10 m/sec. The accuracy of position measurements of a ball on the table is about 1 mm. When travelling with the maximum speed, it takes a ball about 100 μsec to cross 1 mm. We thus assume that a time-slice with a duration of 100 μsec is appropriate to model this application. The continuous movement of a ball in the target is thus expressed in the model by a discrete movement of the ball within 100 μsec steps. The discretization error is then in the same order of magnitude or smaller than the unavoidable measurement error.

If a ball rolls slowly, e.g., 0.01 m/sec, then within a time-slice of 100 μsec the ball will only advance by 1 μm—although such a small distance cannot be measured in our setup, it must be considered in the model in order to avoid an accumulation of round-off errors in a model that implements receding horizon control. In the model we thus propose to express the distance in increments of 1 μm.

Let us now look into the details of what is happening when two elastic billiard balls collide. The assumption, that a change of direction occurs *instantaneously*, i.e., within an *infinitesimal* small time interval, cannot be correct, since this would imply an *infinitely large* acceleration. If we go into the details of an elastic collision between two balls we learn that the mechanic wave propagation inside a ball and Hook's law must be considered. In order to construct such a detailed model of what is happening during a collision, we need a very fine granularity of our digital time scale.

However, on a higher level of abstraction the *law of conservation of momentum* during the collision of two elastic bodies suffices to explain the phenomenon without a need to consider what happens inside an elastic ball at this finer time scale. Whenever a ball hits another ball, the momentum of the interacting balls at the beginning of the collision must be the same as the momentum of the interacting balls at the end of the collision. We can thus assume that the collision takes place within a time-slice with the starting state of the collision at the beginning of this time-slice and the terminating state of the collision at the end of this time slice. By introducing a proper *higher level of conceptualization* (i.e. the *law of conversation of momentum* in the case of billiard ball collisions) and a fitting duration of a time slice we have drastically simplified the analysis of the collision problem.

This technique of simplification by building a hierarchy of models where the relevant characteristics of a complex low-level phenomenon are properly conceptualized by a much simpler high-level concept is fundamental to the process of simplification and to our understanding of the world.

4.6   Assumption Coverage Versus Model Complexity

During the construction of model of a *target in the physical world—a natural system*—one must make a set of assumptions about which ones of the innumerable properties of a natural system must be part of the model and which properties can be disregarded when constructing the model for the stated purpose. In Sect. 2.3 we introduced the phrase *ceteris paribus* to denote all those properties that are regarded irrelevant in the process of model construction, since these properties are considered *constant* in the Interval of Discourse (IoD) and thus have no detectable effect on the observable behavior in the IoD.

The term *assumption coverage* [Pow95] refers to the probability that all assumptions that are made during model construction are met by reality. In an open system, the assumption coverage will always be less than one.

Example 4.6.1:   if we assume that the temperature of an ideal gas is constant, then the simple relation can express the relation between volume and pressure of the gas:

$$\text{Volume} \times \text{Pressure} = \text{Constant\_1} \qquad\qquad [1]$$

If we discard the assumption *constant temperature* during the IoD, then the relation reads

$$\text{Volume} \times \text{Pressure} = \text{Temperature} \times \text{Constant\_2} \qquad\qquad [2]$$

It is evident, that relation [2] has a higher *assumption coverage* than relation [1] at the expense of a higher complexity of the model.

This relation between *assumption coverage* and *model complexity* is important from the point of view of complexity. An increase in the *assumption coverage* is often realized by an accompanied increase in the *model complexity*, because more properties of the natural system are included in the model.

Two models are *compatible*, if one model is a generalization of the other model. In example 4.6.1 the model that implements relation [2] is thus *compatible* with the model that implements relation [1].

There are many assumptions that cannot be properly included in the model of a natural system. Consider the example 2.3.4 about an *earthquake* occurring during a game of billiards. It is impossible to anticipate the exact instant and the precise characteristics of the earthquake ahead of the time of its occurrence.

Example 4.6.2: When designing a model for an engineering artifact, e.g. a dam against a tidal wave of a tsunami, one assumes about the maximum size of the tidal wave based on observation of the ocean during the last relevant period, e.g., the last 100 years. The *assumption coverage* that the dam will withstand a given flood is thus limited by the probability that a flood larger than the assumed flood will occur.

In a number of situations, the diverse facets of a single natural phenomenon cannot be explained in a single model or in compatible models.

We call two models *complementary* if they explain different facets of a single natural phenomenon and if they are not compatible.

Example 4.6.3: A quantic entity, e.g., the behavior of light, can be partly modeled by a classical wave model and partly by a classical particle model. These incompatible but complementary models are needed to capture the full nature of the quantic entity. This wave-particle duality is thus a property of the models.

4.7 Points to Remember

- A *model of a system* is a simplified representation of the system (the *target*) in order to understand the *target* for a given purpose. *Models of reality* are always *idealizations*, i.e. deliberate *abstractions* that make simplifying assumptions in order to make *reality* tractable.
- A *target* in the physical world must obey the natural laws, e.g., the laws of physics. A *model* resides in a symbol world, follows the rules of this symbol world and imitates the behavior of the *target* by symbol manipulation. The *modeling relation* is the relation between a *target* and the *model of the target.*
- The *principle of parsimony* of William Ockham states that whenever different models can explain a set of observations the model that relies on the smallest set of assumptions—the *simplest hypothesis*—wins.
- The term *assumption coverage* refers to the probability that all assumptions that are made during model construction are met by reality. In an open system, the assumption coverage will always be less than one.
- We call the concrete steps of *symbol manipulation,* prescribed by a *model of behavior* of the target, that lead from a *start state* to an *end state* in the symbol world the *model algorithm.* A *model algorithm per se* is timeless, although the execution of the algorithm, e.g. on a given digital computer, takes a certain real time-interval—the *execution time of the algorithm.*
- In *receding horizon control* the *arrow of real-time* is partitioned into a sequence of equidistant time-slices, the *sampling interval*s. At the end of a time-slice the prediction horizon is shifted forward towards the future.

- The physical duration of the time slice must be known *a priori*, because it is the important input to the model algorithm that determines the point in time for which the predictions of the model must be made.
- A *mental model* is a model in the conceptual landscape of a human that helps a person to orient herself in the world. Humans have a multitude of mental models of their own behavior and of the perceived structure and behavior of their environment in their minds.
- *Rational models* are part of the *rational system* of the human mind and require an *explicit cognitive effort* for their execution. *Scientific models* are a special class of rational models, where all assumptions are explicitly listed and any element of subjective input is avoided.
- There are two basic techniques available for scientific model construction, the *input-output technique* and the *state variable technique.*
- A *pragmatic mental model* (PMM) for decision making in a real-life scenario is composed of a combination of recognized causal dependencies among the perceived events (using the *rational system*) and an intuitive, sometimes emotional, judgment derived from past experiences (retrieved from the *intuitive system*).
- A *cyber model* is a model for a digital computer to predict or control a process in the physical environment. Since a human designer is required to design a cyber model, there is no *cyber model* without a prior *mental model* in the mind of the designer.
- The model of time in a digital computer is *discrete*—time progresses in steps. Neither *continuous* nor *instantaneous* behavior can be precisely expressed with a *discrete model of time.*
- In order to achieve *human understanding* and explanatory power, it is often required to develop a *new conceptualization* that aggregates the essential elements of a complex, low-level model into a new *single unit of thought.* This technique of simplification by building a hierarchy of models where the relevant characteristics of a complex low-level phenomenon are properly conceptualized by a much simpler high-level concept is fundamental to the process of simplification and to our understanding of the world.

Chapter 5
Multi-Level Hierarchies

"If there are important systems in the world that are complex without being hierarchic, they may to a considerable degree escape our observation or understanding."

—Herbert Simon.

5.1 Introduction

We find an immense number of different things in the world—inorganic things and living things. Classical physics tell us that all these things are composed of only three basic entities: protons, neutrons and electrons. How does the enormous diversity of things get into the world if there are only three basic entities? The answer to this question is: emergence caused by the differences in organization, i.e., differences in the arrangement of and interactions among the parts of a whole.

Example 5.1.1: In 1913 Niels Bohr proposed a simple model that explains how atomic elements are organized by the different arrangements of the three basic building blocks: protons, neutrons and electrons. According to the Bohr model, an atom contains a positively charged nucleus, consisting of protons and neutrons, surrounded by a shell of negatively charged electrons. Electric forces hold the nucleus and the electrons together. The number of electrons in the shell that equals the number of protons in the nucleus determines the basic properties of an atomic element. At a higher level any one of the fewer than one hundred stable atomic elements can now be seen as a basic building block.

We are now in the position to conceive of many more novel structures by considering the different arrangements of the stable atomic elements—the parts. We enter the domain of chemistry. The number of different arrangements of interacting stable atomic elements that lead to a higher-level entity—a molecule—is beyond comprehension.

If we now consider a molecule as a basic building block we arrive at another higher-level of organization—the domain of biology that deals with organisms.

In the preceding paragraphs we have introduced an organizational structure to represent description models of reality at three identified levels: (i) the atomic level

© Springer Nature Switzerland AG 2019
H. Kopetz, *Simplicity is Complex*, https://doi.org/10.1007/978-3-030-20411-2_5

that considers how stable atomic elements are formed by the arrangement and inter-actions of the three basic entities proton, neutron and electron, (ii) the molecular level that considers how molecules are formed by the arrangement and interactions of stable atomic elements, and (iii) the biological level that considers how organism are formed by the arrangement and interactions of molecules. We call such an orga-nizational structure that describes reality at a number of identifiable levels a multi-level hierarchy.

It is the purpose of this chapter to explain how a diversity of novel phenomena can emerge by the integration of models in a multi-level hierarchy.

5.2 Hierarchy Levels

The above-sketched multi-level hierarchy can be extended downwards and upwards. A downward extension could introduce a new level that focuses on the interactions of elementary particles that form an atomic nucleus. The upward extension could introduce new levels that describe organs, plants, animals, humans, families, societ-ies and so on [Mat92].

A level of a multi-level hierarchy is characterized by a set of level-specific entities with distinctive properties and a specific set of interactions among these entities that are controlled by a set of level-specific rules or laws.

Whereas at the lower levels of the above-sketched multi-level hierarchy physical forces (e.g., electromagnetic forces) determine the interactions among the entities, at higher levels of this hierarchy the exchange of information items (Itoms) governs the interactions among the entities.

The physical interactions can be classified in the following three dimensions: (i) distance among the parts, (ii) force fields among the parts and (iii) frequency of inter-actions among the parts. In general, as we move up the levels of a physical hierarchy the distance increases, the force-field magnitude decreases and the frequency of interactions decreases. With every level of a multi-level physical hierarchy a charac-teristic grain of observation can be associated. Phenomena that change much faster than this level-characteristic grain of observation cannot be observed at this level, while phenomena that change much slower than the level-characteristic grain of observation can be considered constant in the selected interval of discourse (IoD).

A multi-level hierarchy is a recursive modeling structure where a system, the whole at the level of interest (the macro-level), can be taken apart into a set of sub-systems, the parts, that interact statically or dynamically at the level below (the micro-level). The properties of the whole are not only determined by the properties of the parts but also by the organization and the interaction patterns among the parts. In many cases, the properties of the whole are significantly different from the properties of its parts.

Example 5.2.1: Under prevailing environment conditions, water—the whole— is a stable compound that extinguishes a fire, while oxygen and hydrogen, the parts of water, boost a fire.

Each one of these parts, the sub-systems, can be viewed as a whole of its own when the focus of observation is shifted from the level above to the level below. This recursive decomposition ends when the internals of a sub-system are of no further interest. We call a sub-system, or a part, the internals of which are of no interest at the current level of abstraction a component (see also Sect. 11.3).

If we model the behavior of a target along the above-sketched multi-level hierarchy, the modeling relation introduced in Chap. 4 suggests that a hierarchy of models must be provided.

A multi-level hierarchy is a blueprint for the integration of a multitude of different models where each one of the models represents and explains an aspect of the target. Recall the beginning of Chap. 4: *A model of a phenomenon is a simplified representation of the phenomenon that is built by humans in order to understand a given phenomenon for a given purpose.* Each one of the models of a multi-level hierarchy should be simple enough that it is amenable to human understanding.

5.2.1 Holons

Arthur Koestler has introduced the term *holon* [Koe67, p. 341] to refer to the *two-faced character* of an entity in a multi-level hierarchy. The word holon is a combination of the Greek "holos", meaning all, and the suffix "on" which means part. Figure 5.1 depicts a graphical representation of seven holons. The point of view of the observer determines which view of a given holon is appropriate in a particular scenario.

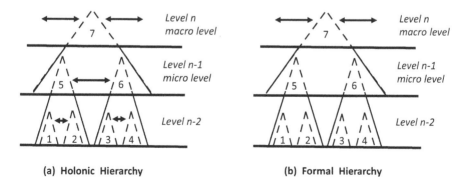

(a) Holonic Hierarchy (b) Formal Hierarchy

Fig. 5.1 Holonic Hierarchy (**a**) and Formal Hierarchy (**b**)

Viewed at the macro-level, *level n,* the whole—holon 7 in Fig. 5.1— is a stable entity that encapsulates and hides its parts, i.e., the holons 1–6 that form the whole by interacting at the lower levels.

This stable entity—the whole—often requires a new conceptualization and its interactions with other wholes at the macro-level give rise to novel laws.

At a given macro-level, we consider the whole (the stable entity) surrounded by a skin. Interfaces located at the skin of this stable entity allow the exchange of matter, energy or information to realize the interactions among wholes at the same level or of the whole with its environment.

Viewed at the micro-level—*level n-1*—, a holon is characterized by a set of interacting parts—the components— (holon 5 and holon 6 of Fig. 5.1) that are confined by the skin of the holon.

If we shift our attention from *level n* to *level n-1,* then level *n-1* becomes the macro-level and *level n-2* becomes the micro-level. At the new micro-level, *level n-2,* the parts of a whole at the new macro *level n-1* become visible.

Figure 5.1 (a) shows that at any given level all interactions that occur within lower levels of a holon are enclosed by the skin of the holon. This leads us to the following *holonic enclosure restriction*:

The parts of a holon at the micro-level may only interact with other parts of the same holon and must not interact with entities outside of the holon.

This holonic enclosure restriction that restricts the interactions of a holon at the micro-level is absolutely essential to maintain the integrity of the abstraction of a holon as a whole at the macro level.

We call a multi-level hierarchy where the holonic enclosure restriction is satisfied at every level a **holonic multi-level hierarchy**.

The holonic enclosure restriction provides a guideline for the design of interfaces. The strict rules for the interfaces to the external environment in a holonic multi-level hierarchy support the simplification by abstraction. At any given level of a holonic multi-level hierarchy the understanding of the function of the holon at that level is sufficient to integrate the holon with other holons at that level. There is no need to understand the lower-level interactions of the parts within a holonic multi-level hierarchy.

A holon is a model and the holonic enclosure restriction is a *modeling restriction*. This modeling restriction is necessary to simplify the reasoning about the behavior of a holon at the macro-level and to forget about all issues that occur at the micro-level or below.

There are physical systems that do not fully comply with the holonic enclosure restriction, as shown in the following example 5.2.2.

Example 5.2.2: Consider the interactions that are caused by a radioactive atom in one of the molecules of an organism. These interactions violate the holonic enclosure restriction.

This example shows that it is hardly possible to develop a holonic multi-level hierarchy of models of a physical entity that deals with every facet of physical reality.

However, in many situations a holonic multi-level model of some important facet of a system can help to explain the behavior of the system *ceteris paribus* (see Sect. 2.3).

Whenever we build a large artifact—e.g., a large computer system— and structure the artifact in the framework of a multi-level hierarchy we should try to comply—whenever possible—with the holonic enclosure restriction such that the resulting hierarchy becomes a holonic multi-level hierarchy.

Let us summarize again the essence of this important holonic enclosure restriction (see also example 5.3.4 in this Section):

- At the micro-level a holon is an isolated unit with no interactions outside the scope of the holon.
- All interactions of a holon with the world outside the holon, i.e., to other holons or the environment, must be lifted up to the macro-level.

5.3 Hierarchy Types

There can be many different relations between the whole at the macro level and parts at the micro-level as well as among the parts at the micro-level. Examples of relations are

- Material hierarchy: the whole comprises the parts
- Uses hierarchy: the whole uses the parts
- Description hierarchy: the whole can be described at different levels of abstraction
- Information hierarchy: the whole interacts with the parts by the exchange of Itoms
- Control hierarchy: the whole controls the parts

These relations are by no means exclusive. For example, a material hierarchy is also a description hierarchy. The following example 5.2.1, taken from Ahl [Ahl96, p. 107] shows the fine difference between a material hierarchy and a control hierarchy.

Example 5.3.1: The concept army denotes a material hierarchy that consists of the soldiers of all ranks and contains them all. In contrast, a general at the top of an army (a military hierarchy) controls the soldiers, but does not contain them. In this example of the military hierarchy the control constraints originate from outside, i.e. above the macro-level.

In the following we discuss three types of multi-level hierarchies more closely: a *formal hierarchy,*- an *information hierarchy,* and a *control hierarchy.*

5.3.1   Formal Hierarchy

A hierarchy is a formal hierarchy if there are no interactions among the parts at the micro-level, i.e., the parts are only related to the whole of the macro- level.

If the parallel lines within *level n-1* and *level n-2* of Fig. 5.1 (a) are deleted, we have the picture Fig. 5.1(b) of a formal hierarchy.

Example 5.3.2:   The soldiers in a military organization that relate at a given level only to their boss, but not to each other, establish a formal hierarchy.

A formal hierarchy is the simplest form of a hierarchy, since there are no horizontal interactions among the parts at the lower levels. Each part of a formal hierarchy can be investigated in isolation from the other parts (see Sect. 9.2 on divide and conquer). A formal hierarchy thus supports the side-effect free composition of the parts, where composability is defined as follows [Kop97, p.34]:

"An architecture is composable with respect to a specified property if the system integration will not invalidate this property once the property has been established at the subsystem level."

Composability of properties is a desired characteristic of an architecture, because it supports the orderly integration of the parts.

5.3.2   Information Hierarchy

A multi-level hierarchy where the interactions among the entities (e.g., the parts of a distributed computer system) are realized by the exchange of Itoms is an information hierarchy.

In cyberspace, an information hierarchy is the dominant organizational framework for structuring the interactions among the components of a distributed computer system. The interactions within an information hierarchy are realized by the exchange of messages that transport the data parts of the Itoms. From the point of view of semantics, these message exchanges establish the *degree of dependence (DoD)* among the components at a semantic level.

The degree of dependence (DoD) denotes to what extent the function of one component depends on the correct operation of other interacting components.

In a large distributed system where the components interact by the exchange of Itoms, the components should possess a certain level of autonomy—i.e., the components should be loosely coupled to reduce the degree of dependence. An autonomous component can often take an alternate path of action if a partner does not perform due to an error or for some other reason. If a component is fail-silent—it produces either correct messages or no messages at all (see Sect. 11.4)—then an error in the message transmission can only be detected in the temporal domain by the receiver. In a Cyber-Physical System, where reaction time is constrained, the error-detection latency must be tightly bounded in order that an alternate action to mitigate the consequences of an error can take place within the limited time window. A small error-detection latency can only be achieved if the specified temporal framework of the interactions is tight and known to the receiver. It follows that a tight temporal framework does not necessarily imply a high DoD in a Cyber-Physical System.

> *On the contrary, a tight temporal framework can help to improve the autonomy of the parts because communication errors are immediately detected and the effect of these errors can be handled in time.*

A unidirectional *message exchange* is an asymmetrical interaction between a sender and a receiver. The correct operation of the receiver of a message depends on the correct operation of the sender that produces the expected message, whereas the correct operation of the sender does not depend on the correct operation of the one or many recipients of the messages. This unidirectionality of a message exchange reduces the coupling—the DoD—of the components. It is only kept up in the architecture if the transport protocols that implement a message transport at the lower levels do not require acknowledgement messages that travel in the other direction, i.e. from the recipients of a message back to the sender.

Example 5.3.3:   If a unidirectional information flow among components is required at the application level, one should not use a communication protocol that demands a bidirectional communication link at the implementation level (e.g., a low-level *Positive Acknowledgement* or *Retransmission Protocol*). In a unidirectional information flow, the behavior of the sender does not depend on the behavior of the receiver. If a bidirectional communication protocol is deployed, an unnecessary additional dependency of the sender on the behavior of the receiver is introduced.

The above arguments concerning the capability to detect communication errors immediately by the receiver of a message and to preserve the unidirectionality of the message transfer throughout the architecture have provided one important rational for the design and use of the time-triggered architecture and the time-triggered protocols in Cyber-Physical Systems (see also Sect. 8.3).

5.3.3   Control Hierarchy

A control hierarchy can be seen as a special case of an information hierarchy, where the exchanged messages contain control commands to the lower levels. In a control hierarchy the macro-level provides some constraints on the behavior of the parts at the micro-level, thus establishing a causal link from the macro level to the micro-level. These constraints, called *downward causation,* restrict the behavior of parts beyond the natural laws, which physical parts must always obey.

In a hierarchical organization the power to influence the course of events—the control—percolates from the top of the hierarchy downwards. In some cases, the control constraints can have their origin inside the whole, i.e. the collective behavior of the parts at the micro-level. Since behavior is a concept that depends on the progression of time, there is a temporal dimension in control hierarchies that deal with behavior.

Whenever the behavior of the parts at the micro-level causes the behavior of the whole at the macro-level and the whole constrains the behavior of the parts, we have upward causation from the parts to the whole and downward causation from the whole to the parts.

Example 5.3.4:   The development of a traffic jam on a busy highway. The movement of the cars causes a traffic jam while, at the same time, a traffic jam causes a delay in the movement of the cars.

Downward Causation   The interactions of the parts at the micro-level cause the whole at the macro-level while the whole at the macro-level constrains the behavior of the parts at the micro-level. *"Downward causation is a difficult concept to define precisely, because it describes the collective, concurrent, distributed behavior at the system level .... Downward causation is ubiquitous and occurs continuously at all levels, but it is usually ignored simply because it is not under our control .... The motion of one body in an n-body model might be seen as a case of downward causation* [Pat00, p. 64]." Downward causation can establish a *causal loop*.

Pattee [Pat73] discusses control hierarchies extensively in *The Physical Basis and the Origins of Hierarchical Control*. In order to support the simplification at the macro-level, a control hierarchy must on one side *abstract from* some degrees of freedom of the behavior of the parts at the micro-level but on the other side must *constrain* some other degrees of freedom of the behavior of the parts, i.e., a control hierarchy must provide constraints from above, while, in a multi-level material hierarchy the natural laws provide constraints from below.

The delicate borderline between the constraints from above on the behavior of the parts at the micro-level and the freedom of behavior of the parts at the micro-level is decisive for the proper functioning of any control hierarchy.

There are two extremes of control which lead to a collapse of the control hierarchy: (i) full control from above which defeats the principle of abstraction of control and (ii) no constraints from above which can lead to unconstrained chaotic behavior.

Example 5.3.5:   A good conductor of an orchestra will control the tempo of the performance without taking away the freedom from the musicians to express their individual interpretation of the music.

5.4 Emergence

It is quite common, as we move up a multi-level hierarchy, that novel phenomena can be observed at a given level, that are not present at the level below. We call these novel phenomena *emergent phenomena*. We use the term *phenomenon* as an umbrella term that can refer to structure, property or behavior.

"A phenomenon of a whole at the macro-level is emergent if and only if it is of a new kind with respect to the non-relational phenomena of any of its proper parts at the micro level [Kop16, p.82]."

A phenomenon is of a new kind if the concepts that describe this phenomenon cannot be found in the world of the isolated parts. *Conceptual novelty* is thus the landmark of our definition of emergence. Note that, according to the above definition, the emergent phenomena must only be of a new kind with respect to the non-relational phenomena of the parts, not with respect to the knowledge of the observer. If a phenomenon of a whole at the macro-level is not of a new kind with respect to the non-relational phenomena of any of its proper parts considered in isolation at the micro level then we call this phenomenon *resultant*.

Take the example 5.2.1 of water. Both oxygen and hydrogen, the constituent parts of water, boost a fire, while water extinguishes a fire. The property of water to extinguish a fire is thus an emergent property that is novel with respect to. the properties of the parts—oxygen and hydrogen—which both boost a fire. The reductionist view of the world that says that every complex phenomenon can be fully explained by analyzing the properties of its constituent parts in isolation reaches its limits when emergence is at work.

The description of an emergent phenomenon often requires the formation of a new concept—a new unit of thought—to capture those aspects of the interactions of the parts at the micro-level that are of relevance for the purpose of the multi-level modeling structure at the macro level.

The incorporation of all those aspects of the interaction of the parts at the micro-level that are considered relevant for the macro-level into a unified new concept at the macro-level can lead to an abrupt simplification at the interface between the micro-level and the macro-level of a multi-level hierarchy.

It is often the case that new regularities are observed at the macro level that are not present at the micro level.

Example 5.4.1:   *Stoke's law* about the behavior of a spherical object in a fluid of particles is only relevant at the macro-level, where a fluid of particles exists, but not at the micro-level, where each particle is considered in isolation.

In many cases of emergent behavior, we find a control hierarchy with downward causation where a causal loop exists between the macro-level and the micro-level. We can observe such a causal loop in many scenarios that are classified as emergent in every-day language: the behavior of birds in flocks, the synchronized oscillations of fireflies or the build-up of a traffic jam at a congested highway.

Example 5.4.2:   The following example of distributed fault-tolerant clock synchronization shows how the novel notion of a *fault-tolerant global time* emerges in a control hierarchy with a causal loop out of a set of non-fault-tolerant clocks.

A distributed fault-tolerant synchronization algorithm comprises the following three phases [Kop11, p. 69]:

- Periodic exchange of the time value of the local clock of each computing node among all the nodes of the system.
- Distributed calculation of a global fault-tolerant time value, taking the local readings of the clock as inputs.
- Adjustment of the local clock to come into agreement with the calculated global fault tolerant time value.

According to the theory of clock synchronization the number N of physical clocks in a system must be larger than $3\ k$, where k is the number of faulty clocks i.e., $N \geq (3\ k + 1)$. A physical clock is a device that contains a physical oscillator (e.g., a crystal) and a counter that counts the number of ticks of the oscillator and thus contains the state of the clock. The frequency of the physical oscillator is determined by the laws of physics and depends on the size of the crystal and environmental conditions, such as temperature or pressure—a case of upward causation. The speed of the oscillator cannot be modified by downward causation. However, the state of the clock is modified by downward causation in step three of the algorithm. The phenomenon fault-tolerant clock synchronization fulfills the requirement of an emergent phenomenon:

- The phenomenon fault-tolerant time, which does not fail if a single clock fails, is novel with respect to the behavior of a single clock that can fail.
- There is downward causation. The system of concurrently executing clocks constrains the execution of an individual clock by adjusting the state of the counter of the local clock to a value that has been determined by the ensemble of clocks.

This example of emergence is interesting from the point of view of how upward causation (the frequency of a physical clock) and downward causation (the periodic correction of the state of a clock caused by the time value calculated by the ensemble of clocks at the macro level) interact and form a causal loop.

Emergent Properties can be classified according to Fig. 5.1. Most problematic are the emergent properties that have a negative effect and are unknown at design time.

There are *phenomena of emergence* that cannot be explained within today's state of science. An example is the phenomenon of life or the emergence of the mind over the neurons of the brain.

Fig. 5.1 Classification of emergent properties

Effects of Emergence	Emergent Properties	
	Known-Planned	Unknown-Surprise
Positive	Outcome of Design	Positive Surprise
Negative	Avoided by Design	Impedment

5.4.1   Supervenience

In a multi-level hierarchy, the novel properties of the macro-level are formed by the properties and the interactions (the organization) of the parts at the micro-level. We say that the macro-level properties *supervene* on the micro-level properties. According to this view, chemical properties supervene on physical properties and biological properties supervene on chemical properties.

Example 5.4.3:   Human thought supervenes on the biological activities of neurons.

The principle of *supervenience* establishes two important dependence relations between the emerging phenomena at the macro-level and the interactions and organization of the parts at the micro-level. Supervenience can be broken down into two parts, the *Abstraction Relation* and the *Diagnosis Relation*.

Abstraction Relation   A given emerging phenomenon at the macro level can emerge out of many different arrangements or interactions of the parts at the micro-level. Because of this abstraction relation one can abstract from the many different arrangements or interactions of the parts at the micro level that lead to the same emerging phenomena at the macro level—a very powerful simplification.

Example 5.4.4:   A partly random process determines the intended distribution of the dopants in a semiconductor crystal. It is highly probable that in a VLSI chip with 1 million transistors, the microstructure of every single transistor is different from the microstructure of every other transistor. As long as the electrical characteristics of a transistor are within the specified range, these differences at the micro-level are of no relevance for the user of a transistor. The abstraction relation thus entails a significant simplification of the higher-level models of this multi-level hierarchy since—despite their individual differences—all transistors can be treated alike.

Diagnosis Relation   A difference in the emerging phenomena at the macro- level requires a significant difference in the arrangements or the interactions of the parts at the micro-level. Because of this diagnosis relation any difference in emerging phenomena at the macro level can be traced to some significant difference at the micro level. The diagnosis relation is important from the point of view of failure diagnosis.

Example 5.4.5:   If a transistor does not exhibit the specified behavior at the macro-level, then there must be some significant difference in the arrangement of the dopants at the micro-level.

The question, whether the properties of the macro-level can in general be reduced to the properties of the parts at the micro-level is a topic of an intense philosophical debate. The properties of the parts at the micro-level together with the organization and interaction of the parts at the micro-level determine the properties of the whole of the macro level. This is definitely more than the properties of the parts of the micro-level considered in isolation.

At any level of a multi-level hierarchy the model at that level—we call it the *level model*— can be characterized by the concepts describing the entities and the interactions among these entities at that level. The *enclosure property* of a holon guarantees that the activities at the micro-levels are encapsulated and confined to the internals of a holon. All external interactions of a holon occur at the macro level. The (possibly novel) concept of the holon at the macro-level encompasses all those aspects of the holon that are relevant for the given purpose of the macro-level. There is thus no need to consider the processes that occur at the micro-levels inside the holon.

Simon [Sim69, p. 209] argues that in many systems the laws and rules that govern the behavior at a level are nearly independent of the level above and the level below, giving rise to the principle of *near decomposability* of levels.

This principle of near decomposability states that an approximate model suffices in most cases to model the behavior at any given level of a multi-level hierarchy.

This approximate model—the level model—considers only the interactions at the considered level and abstracts from the behavior of the high-frequency parts at the level below and considers the parameters of the low frequency parts at the level above as constants.

5.5   Points to Remember

- A *multi-level hierarchy* is a recursive modeling structure where a system, the *whole* at the level of interest (the *macro-level*), can be *taken apart* into a set of sub-systems, the *parts*, that *interact* statically or dynamically at the level below (the *micro-level*).
- A level of a multi-level hierarchy is characterized by a set of distinctive entities with unique properties and a set of level-specific rules or laws that govern the interactions among these entities.
- The properties of the whole—at the macro level—are not only determined by the properties of the parts at the micro-level but, more importantly, by the interaction patterns among the parts. In many cases, the properties of the whole are significantly different from the properties of its parts.
- A multi-level hierarchy where the interactions among the entities (e.g., the components of a distributed computer system) are realized by the exchange of Itoms is an *information hierarchy*.
- The degree of dependence (DoD) denotes to what extent the function of one component depends on the correct operation of other interacting components.
- A hierarchy is a *formal hierarchy* if there are no interactions among the parts at the micro-level, i.e., the parts are only related to the whole of the macro- level.
- Arthur Koestler introduces the term *holon* for the *two-faced character* of an entity in a multi-level hierarchy. Viewed at the macro-level, the holon is a *stable entity* that encapsulates its parts and provides a service to its environment. Viewed at the micro-level, a holon is characterized by a set of interacting parts

that are encapsulated by the skin of the holon, giving rise to the *holonic enclosure restriction.*

- A hierarchy is a *holonic hierarchy* if the parts of a holon at the micro-level may only interact with other parts of the same holon and must not interact with entities outside of the holon.
- Downward causation is a control relation between the whole and the parts. Downward causation can establish a *causal loop* between the *micro-level* and the adjacent *macro level.*
- A phenomenon of a whole at the macro-level is *emergent* if and only if it is of a new kind with respect to the non-relational phenomena of any of its proper parts at the micro level.
- The principle of *supervenience* establishes the *Abstraction Relation* and the *Diagnosis Relation* between the emerging phenomena at the macro-level and the interactions and arrangement of the parts at the micro-level.
- This principle of *near decomposability* states that an *approximate* model suffices in most cases to model the behavior at any given level of a multi-level hierarchy.
- The incorporation of all those aspects of the interaction of the parts at the micro-level that are relevant for the macro-level into a *unified new concept* at the macro-level can lead to an abrupt simplification at the interface between the micro-level and the macro-level of a multi-level hierarchy.

Chapter 6
Cyber-Physical Systems Are Different

"Once cyber crosses into the realm of the physical, then it's a physical attack, but it starts with cyber. And the idea of a cyber-attack being able to take control of machines—that becomes a scary process."

—David E. Sanger

6.1 Introduction

Cyber-Physical Systems differ from *conventional data-processing systems* in a number of important aspects. These differences are so profound that they require a new state of the mind when dealing with Cyber-Physical Systems [Lee08].

A conventional data-processing system is a *symbol-manipulation system* that operates solely in cyber space and interacts with an intelligent human being in the physical world across a human-machine interface at instants that are decided by the computer. A human-machine interface accepts the input data from a human operator and delivers the output data to this human operator. The human operator checks these outputs for plausibility before he uses them to initiate an action in the physical world.

A Cyber-Physical System consists of two subsystems, the physical process in the environment and the controlling computer system (the cyber system) that observes selected state variables of this process and controls this process by providing set-points to actuators that change values of quantities in the physical world. Often there is no human in the loop who observes the environment with his diverse human sense organs, develops a mental model of the ongoing activities, and provides an *intelligent buffer* between cyber space and physical space.

It is the purpose of this chapter to explain in detail the differences between a conventional data processing system and a Cyber-Physical System (CPS).

© Springer Nature Switzerland AG 2019
H. Kopetz, *Simplicity is Complex*, https://doi.org/10.1007/978-3-030-20411-2_6

6.2   Autonomous Operation in the Physical World

The first and most distinctive difference between a conventional data processing system and a CPS relates to the operation of a CPS that has the capability to autonomously modify the state of the physical world. The available sensors of the cyber system determine and limit the view that the cyber system has of the physical world. The cyber model (see Sect. 4.5) that controls the processes in the physical world is thus constrained by this limited view and cannot detect changing properties of the physical environment caused by environmental dynamics that are outside the scope of the sensors provided by the system designer. An unforeseen environmental event or a failure of a single senor can cause a serious malfunction of a CPS [Tör18]. This is in contrast to a human operator with his diversity of redundant biological sensors and a large intuitive knowledge base in his conceptual landscape.

> **Example 6.2.1:** A human driver of an automobile sees the traffic on the road, hears an approaching ambulance, feels the bumps on the road and smells the overheated brake.

A human operator can detect abnormal situation and edge-cases that have not been anticipated by the human designer of the cyber system. (An *edge case* is an operational situation in the physical environment that occurs with a very low probability).

> **Example 6.2.2:** There are many more edge cases that must be properly handled by an autonomous car when driving within a city than when driving on a freeway.

As long as a malfunction of an autonomous system has limited impact on the physical environment, a failure of the CPS is non-critical. However, many CPS are controlling environments, where a failure of the cyber system can have catastrophic consequences in the physical world.

6.3   Timeliness

Cyber-Physical Systems have to be aware of the progression of physical time in the physical world. This has severe consequences for the system design in the following areas:

- Pacing
- Execution Time of Algorithms
- Simultaneity

6.3.1 Pacing

In a conventional data processing system, the pacing of the dialog between the computer and the human operator is performed by the computer: after delivering an input, the human operator has to wait some physical time—we call it the *response latency*—until the computer-system provides the corresponding output. The computer extends this response latency if the workload of the computer is increased. If the response latency reaches a value beyond a reasonable limit or if the result delivered by the computer seems implausible to the human, then the human user considers the computer system to have failed and takes some mitigating action. Thus error-detection and error handling are in the responsibility of the human operator.

In contrast to this, the process in the physical environment determines the pace of a CPS. The progress of the physical process, determined by the progression of the physical time, dictates at what instant on the physical time line an action by the controlling computer system must take place. The outputs of a CPS must thus be correct in the value domain and in the temporal domain. In Cyber-Physical Systems the progression of physical time must be considered a *first-order citizen*.

Example 6.3.1 Consider the instant when the fuel must be injected into an automotive engine. A single incorrect timing of the computer output can cause a permanent mechanical damage to the engine.

The correct timing of the CPS outputs must be guaranteed in the full range of operating conditions—also under the rarely occurring condition of peak load. In many applications the utmost utility of a CPS is related to a *predictable peak-load performance*, because in a rarely occurring peak-load situation the services of the computer system are needed most urgently.

Example 6.3.2: Consider the case of a system that controls the electric power grid. If, during a thunderstorm, the power grid is exposed to a number of lighting strokes that cause an avalanche of alarms and the consequences of tripping circuit breakers are not handled appropriately in time by the computer system, then the system fails when it is needed most urgently.

6.3.2 Execution Time of Algorithms

It is a further challenge in the design of a real-time system to develop algorithms that are guaranteed to provide a satisficing result before a deadline that is dictated by the physical process. The two possible approaches to tackle this challenge, the use of WCET (worst-case-execution-time) algorithms or the use of any-time algorithms have been discussed in Sect. 4.2

6.3.3 Simultaneity

The problem of the proper handling of *simultaneous events* that occur in the physical world is one of the tough problems in computer science. It surfaces in the domain of hardware design as the *meta-stability problem*, in the domain of operating design as the *mutual-exclusion problem* and in the domain of distributed system as the *consistent-ordering problem* of messages. In a digital system the exact temporal order of physical events that occur close to each-other cannot be established from the timestamps of the events for two reasons: (i) all events that occur during a single granule of time (i.e. the time interval between two ticks of the available clock) of the digital time-base receive the same time-stamp and (ii) it is impossible to fully synchronize a set of digital clocks in a distributed system. The problem does not disappear if a time-base of finer granularity is chosen [Kop11]. The establishment of a system-wide consistent order of events that occur in a distributed system is of particular importance when fault-tolerance by the replication of deterministic processes is introduced to mask a component failure [Pol96].

6.4 Safety-Criticality

Many Cyber-Physical Systems are safety critical. In a *safety-critical Cyber-Physical System* a failure, caused by a hardware or software fault, can have catastrophic consequences on human life, the environment or the economy.

> **Example 6.4.1:** An example of a safety-critical systems is a control system on-board an airplane or a car, a medical system that interacts directly with the patient or a control system of a nuclear power plant.

> **Example 6.4.2:** Even the correct timing of a simple traffic light is safety critical. A failure of the traffic light to change from green to red at the specified instant will have the same catastrophic consequences as a failure in the value domain.

It is an important characteristic of a safety critical CPS that there is no human in the loop that checks the output of the computer before it is released to the physical environment. The magical number of 10^9 h is often postulated for the mean time between failures (MTBF) of a safety-critical CPS. A single high-quality chip has an MTBF of around 10^7–10^8 h with respect to *permanent failures* and a significantly lower MTBF with respect to *transient failures*. A transient failure of the hardware, caused by a single event upset (SEU), can have a disastrous impact on the function of the software.

> **Example 6.4.3:** Hardware fault-injection experiments in Deep Learning Neural Networks (DNN) have shown that a single bit-flip in the feature classification layer of a DNN can cause a truck on the road to be classified as a flying bird [Li17].

A single thread of execution of complex software on standard (non-fault-tolerant) hardware cannot achieve the high level of reliability required in a safety-critical application.

Based on field experience, Dvorak [Dvo09, p.3] posits that after the *final acceptance test* one residual software error is to be expected in 1000 lines of source code. If a very disciplined software development process is followed then this software error rate can be reduced to one error per 10,000 lines of source code. During normal operation these residual software errors will usually not manifest themselves, since during the operational acceptance tests all residual software errors that had an effect on the normal operation have been eliminated. However, the duration of the operational acceptance test is normally orders of magnitude smaller than the duration of failure free operation (the magical number of 10^9 h) that is required in a safety-critical system. These residual software errors are activated in exceptional cases—the edge cases that have not been fully tested. (See also Sect. 10.6 on *Safety-Critical Software*).

6.5   Cyber Security

The protection of the data and the services of a computer system from malicious actions by an adversary is the concern of *cyber security*. There cannot be safety without cyber security. From the point of view of security, CPSs are different from data-processing systems. Whereas in a data processing system the main concern is the confidentiality and privacy of the data, in Cyber-Physical System the integrity of the actions and authenticity of the information are of utmost importance, while the confidentiality of real-time data is of lesser concern.

A weakness in the design of a system that can be exploited by an adversary in a security attack is called a *vulnerability*. A successful security attack, where an attacker exploits a vulnerability and gains control of the system, is called an *intrusion*. The path that an adversary takes to exploit a vulnerability is called an *intrusion path*. The totality of the vulnerabilities and all possible intrusion paths form the *attack surface* of a system.

Cyber security is an overarching concept. It includes evil actions by a malicious person during any one of the life cycle phases, i.e. the design, operation and maintenance, of a computer system. The result of a security incident can be the installation of *malicious code* (called a *Trojan Horse*) at design time, during a maintenance action, or during normal operation. The Trojan Horse is then activated at a critical instant during the operation of a system at run time. A study on Cyber-Security Incidents [Ogi17] says that insider attacks pose the same level of risk of unauthorized access as remote cyber-attacks.

The first version of many Cyber-Physical Systems is often developed as a stand-alone system without any external connections and little concern about possible vulnerabilities from the Internet. The intrusion path of a stand-alone system requires the physical accesses to the system by an adversary. The physical isolation of the system—e.g., in a fenced area of an industrial enterprise with strict physical access control—is thus sufficient to protect the system from security attacks. As soon as the system is connected by wire-bound or wireless connections to the outside world, the attack surface of a system is substantially increased and remote attacks become possible.

For example, the introduction of wireless channels in modern vehicles increases significantly the attack surface and the range of attack options that are open to a malicious adversary.

Example 6.5.1: Koscher and al. [Kos10] investigated in a systematic analysis the remote security attack surface of a mass-produced passenger car and found out that a malicious adversary may gain complete control over the vehicle's computer systems, including the control of the engine, the steering and the brakes, without ever coming physically close to the vehicle.

It is often assumed that the installation of a *security firewall* is sufficient to protect a system from all security attacks. Experience has shown [Ogi17] that this is not the case. In addition to *perimeter security,* realized by security firewalls, mechanism to detect successful intrusions must be put in place inside a Cyber-Physical System.

In a safety-critical CPS, any security incident can have a direct effect on the safe operation of the CPS, because the consequences of the security incidents can be an unintended modification of the physical environment.

Security attacks against a Cyber-Physical System, particularly those that are executed via the Internet, are thus of utmost safety relevance for safety-critical applications. A good overview of documented cyber-security attacks on *Critical Infrastructure and Industrial Networks* is contained in [Ogi17].

6.6   Points to Remember

- A Cyber-Physical System (CPS) consists of a *cyber-system* that controls a *physical system.*
- A Cyber-Physical System (CPS) has the capability to *autonomously* modify the state of the physical world.
- In a *safety-critical Cyber-Physical System* a failure, caused by a hardware- or a software fault, can have catastrophic consequences on human life, the environment or the economy.
- An output of a CPS is only correct if the *intended value* is delivered at the *specified point in time.*
- The correct timing of the CPS outputs must be guaranteed in the full range of operating conditions—also under the rarely occurring condition of *peak load.*
- The complexity of a computationally expensive algorithm makes it hard or even impossible to establish a WCET (worst case execution time) bound for all data points of the given input domain.

- An *anytime algorithm* guarantees to provide a *satisficing result* before the deadline and continually improves this result until the deadline is reached.
- With a *single thread of complex software executed on standard hardware* the high level of reliability required in a safety-critical application cannot be achieved.
- Cyber Security is an *overarching concept*. It includes evil actions by a malicious person during any one of the life cycle phases, i.e. the design, operation and maintenance, of a computer system.
- In a CPS the *integrity of the actions* and the *authenticity and timeliness of the information* are of utmost importance.
- It is often assumed that the installation of a *security firewall* is sufficient to protect a system from all security attacks. Experience has shown that this is not the case. In addition to *perimeter security,* realized by security firewalls, mechanism to detect successful intrusions must be put in place inside a CPS.
- A study on cyber-security incidents says that *insider security attacks* pose the same level of risk of unauthorized access as remote *cyber-attacks*.
- In a safety-critical CPS, any security incident can have a direct effect on the safe operation of the CPS, because the consequences of the security incidents can be an unintended modification of the physical environment.

Chapter 7
Simplification

"A complex system that works is invariably found to have evolved from a simple system that works. The inverse proposition also appears to be true: A complex system designed from scratch never works and cannot be made to work."

—John Gall.

7.1 Introduction

One fundamental difficulty in gaining an understanding of the behavior of a large artifact is the limited capacity of the cognitive apparatus of humans such that a large monolithic pragmatic mental model (PMM) that describes the full behavior of this artifact is not feasible (see also Sect. 2.5 on the limits of human rational cognition). To arrive at models for human analysis that can be handled by the limited capacity of the human mind we can either divide a large model of behavior of an artifact (divide and conquer) into a number of smaller models—the parts—, to restrict our design to *static structures* or to capture the essence of emergent behavior in a new concept (new conceptualization). In some cases, a combination of these techniques will be most appropriate.

It is the purpose of this chapter to introduce design principles that are intended to simplify a design and to improve the understanding of the behavior of a large cyber-physical computer system. These principles are not fully orthogonal. Each one of them puts the focus on a different aspect of the system.

7.2 Divide and Conquer

The most important simplification technique to reduce the complexity of a large model of a system is the divide and conquer technique that consists of the recursive application of the following steps:

© Springer Nature Switzerland AG 2019
H. Kopetz, *Simplicity is Complex*, https://doi.org/10.1007/978-3-030-20411-2_7

(i)   Partition the large model into a number of simpler and smaller independent parts.
(ii) Arrive at a sub-solution to each one of these simpler parts.
(iii) Combine the sub-solutions to arrive at a combined solution.

The ideas of *partitioning, modular design, separation of concerns, segmentation in the temporal domain* and *some forms of abstraction* are all examples of this divide and conquer technique.

The critical issue in using the divide and conquer technique concerns the interactions among the simpler sub-solutions in the combined solution.

If there are no interactions in the Interval of Discourse (IoD), i.e., the parts are independent of each other, then the combination of the parts is resultant and the recursive application of this divide and conquer principle provides a valid solution. A formal hierarchy (see Sect. 5.2) is an ideal structure for the application of the divide and conquer technique since there are no horizontal interactions within a level, implying that the lower-level parts are only related to the higher-level whole.

Example 7.2.1: In parallel programming the notion of an *embarrassingly parallel algorithm* which requires no communication between the low-level parallel processes is a good example for a formal hierarchy and the application of the divide and conquer technique. These algorithms are widely used in the domain of computer graphics on graphic processing units (GPUs).

Any technique that partitions a large system into two or more (nearly) independent parts is of utmost relevance for the reduction of the system complexity.

Example 7.2.2: Let us analyze at the power of partitioning into independent parts by looking at the number of case distinctions in a software system with sixteen binary decision (Boolean) variables (DV), (see also Fig. 2.1):

1 monolithic system with 16 binary DVs at most: $1 \times 2^{16} = 65536$ cases
2 subsystems with 8 binary DVs each at most: $2 \times 2^{8} = 512$ cases

Since each case requires a careful analysis and a thorough test, the number of case distinctions is a good measure for the objective complexity (see Sect. 2.4 on McCabe's cyclomatic number).

Looking at the *objective complexity measure* of McCabe, the complexity of our system with sixteen Boolean variables can thus be reduced by more than 99% by partitioning the monolithic system into two independent parts. This very significant complexity reduction is the reason why the divide and conquer strategy is of such an importance.

The divide and conquer strategy also opens the possibility to divide the implementation work of a large project into nearly independent sub-projects that can be developed in parallel by different implementation groups without an additional communication overhead. This benefit is important when the system is large and the development time for the new system is constrained.

Observation of Interfaces In most realistic scenarios, the parts will not be fully independent but will interact by the exchange of Itoms across interfaces. During the design the attempt must be made to reduce the degree of dependence (DoD—see Sect. 5.3) among the parts as far as possible and to make the information flow across the interfaces of the parts observable—e.g., by the use of unidirectional multi-cast protocols that send a copy of ever message to a monitor component.

The interactions across the interfaces among the components must be carefully specified, analyzed, and controlled in order to understand the dependencies and to detect potential detrimental emerging phenomena in the combined solution. Particular attention should be paid to stigmergic communication channels that close a loop, since such closed loops are often accompanied by unanticipated emergent phenomena (see Sect. 5.4 on emergence).

As long as the interfaces among the parts remain stable, the internals of the parts can evolve without any effects on the system architecture as a whole.

We can divide a large model of a CPS along the following three orthogonal dimensions

- Partitioning in space (sometimes called *horizontal partitioning*)
- Segmentation in time (sometimes called *step-by-step analysis*)
- Abstraction (sometimes called *vertical partitioning*)

Any separation of concerns that leads to the formation of (nearly) independent parts reduces significantly the complexity of a large system.

7.2.1 Partitioning in Space

There are a number of techniques known to partition a large model into a number of smaller nearly independent models.

Independent Components for Essential Functions If at all possible, every *essential requirement* should be assigned to an *independent top-level component* that implements only this essential requirement and nothing else [Suh90]. The top-level components implementing the essential requirements should be decoupled to the greatest extent possible from each other. A one-to-one relationship between an essential requirement and the corresponding top-level component that implements the associated *essential function* of the evolving system simplifies the:

- *Implementation of the essential function*, since the essential function is decoupled from the implementation of other essential functions. Ideally, each essential function is implemented on a stand-alone top-level component. In a large system, each top-level component is at the root of a multi-level component hierarchy. Care must be taken that the external interactions of this component hierarchy are lifted to the top-level in order that the abstractions of the multi-level component hierarchy remain intact. All external interactions of this component hierarchy must be observable without introducing a probe effect.

- *Evolution of functionality of essential functions*: As operational data about a working system becomes available, it is highly probable that *essential functions* have to be modified. Such a modification is easier if an essential function is implemented in a decoupled stand-alone component.
- *Splitting of the design work*: The design and implementation of a large system requires many persons to work concurrently on the project. If this work can be partitioned among different groups without introducing extra communication-effort, then the time to completion of the project can be significantly reduced.
- *Diagnosis of failures:* If an essential functionality is implemented on a stand-alone component, then the failure diagnosis is straightforward. Ideally the stand-alone component is a Fault-Containment Unit (FCU) such that all internal faults of the stand-alone component—software as well as hardware—are contained within the FCU and the occurring failures can be detected by observing the behavior at the interfaces of the FCU.

Client Actions From Server Actions A server can be seen as a nearly independent subsystem that provides the resources and know-how of the server for the benefit of a client. The normal interface between a client and a server is a message passing communication interface. The partitioning of a large system into clients that take advantage of well-defined services provided by servers implemented in the cloud simplifies the system structure. However, the client must detect the failure of a server promptly in order that the client can take alternative actions to mitigate the consequences of a server failure immediately.

Normal Processing From Safety Assurance In a safety-critical application we need at least three *independent subsystems,* the *normal processing subsystem,* the *safety assurance subsystem* and a *decision system* that decides whether the results produced by the normal processing subsystem are within a *safety envelop* that is established by the safety assurance subsystem. The safety assurance subsystem provides the reference for checking the correctness of the result of the normal processing system. A catastrophic failure can then only occur if both systems, the normal processing subsystem and safety assurance subsystem fail jointly. In order to avoid such a joint failure, the safety assurance subsystem should be completely independent—in hardware and software—from the normal processing subsystem such that the occurrence of common mode failures can be avoided. The software in the decision subsystem that checks whether the results of the normal processing subsystem are within the safety envelope provided by the safety assurance subsystem should be simple enough that it can be implemented with simple software (see Sect. 2.5). The hardware of the decision subsystem that controls the output to the actuators at the interface to the controlled object must be fault-tolerant in order to mask the occurrence of hardware failures (see also the architecture of a safety-critical system explained in Sect. 10.7).

Edge-Case Detection Form Edge-Case Handling   An edge case is an operational situation that occurs with a low probability. A timely edge-case detection must be part of the normal operation; however the edge-case handling should be contained in a separate subsystem that is just concerned with this edge case. This principle helps to reduce the complexity of the subsystem that handles the normal operation. The same arguments apply to the handling of a fault, which can be seen as a special kind of an edge case.

> **Example 7.2.3:** In an autonomous vehicle operation an icy road under foggy conditions at a road construction site is an example of an edge case.

Control Data From Archival Data   In a Cyber-Physical System the handling of real-time control data should be separated from the handling of archival data. Real-time control data has a limited temporal validity and must be processed within give time constraints. The processing of archival data is not time-sensitive; however the long-term reliable storage of archival data on non-volatile storage is more critical than the reliable storage of real-time data.

7.2.2   Segmentation in Time

The principle of segmentation in time by step-by-step analysis segments the behavior of a complex system into a sequence of successive steps on the timeline. The transformations that occur during each step are then investigated in isolation detached from the behavior in the previous and the succeeding step. Finally, the results of this step-by-step analysis are collected to arrive at the complete reconstruction of the behavior of a complex system on the time-line.

The immense advantage of this divide and conquer principle is based on the insight that there cannot be any direct hidden interaction between actions that occur on different sections of the time-line. This principle thus ensures that the parts are independent of each other.

The application of this principle requires that the inputs and outputs of each step are well-defined and can be observed without a *probe effect,* i.e., the observation does not interfere with the execution of a component during a step, neither in the value domain nor in the temporal domain. During the design of a system provisions must be made in order that the principle of step-by-step analysis can be applied. This implies that multi-cast message transmission protocols must be provided such that the periodic messages that are exchanged between the physical world and the cyber world can be routed to an independent monitor system for analysis.

> **Example 7.2.4:** The modeling relation introduced in Sect. 4.2 suggests to segment the time-line into a sequence of equidistant time-slices. If we consider the execution of the control algorithm during a time-slice as a single step, then the state of the control system at the beginning and at the end of each step is well defined. Since the state of the system at the end of a step is identical to the state at the beginning of the following step, it suffices to

observe the state of the system at the *a priori* known periodic sampling instants between two steps. A time-triggered (TT) communication system that supports multicast communication can copy this state into a unidirectional TT-message and can transmit this message to a monitoring system without any side effects (see also Example 5.3.3).

Concurrency that is not properly encapsulated interferes with the step-by-step analysis technique of simplification.

Temporal Domain From Value Domain   In a CPS the control signals that are calculated by the cyber system must be delivered to the actuators at precisely determined instants of physical time. A CPS design is simplified, if the concerns about the CPS behavior in the temporal domain and the CPS behavior in the value domain are separated as far as technically possible.

As outlined in Sect. 4.2 on the modeling relation, the periodic sampling points at the beginning and at the end of a time slice establish a sequence of equidistant time slices, where the end of one time slice coincides with the beginning of the following time slice. At the end of each time slice, a message with the new set points is sent from the cyber world to the physical world and a message with the new observations is sent from the physical world to the cyber world. Within a time slice, time stands quasi still within the model. The duration of the time slice is an important input parameter to the model algorithm since it determines the future instant for which the model predictions must be made.

In a first phase of the design of a time-triggered system, the end points of the time-slices must be determined and fixed on the time axis. The optimal duration between these points depends on the dynamics of the controlled object on one side and the available computing power for the execution of the control algorithms on the other side. Once rough estimates for the duration of the service intervals SI (see Sect. 4.2) are provided, these end points can be fixed. As soon as the estimated sizes of the messages that must be exchanged among the components are available, the development of the model algorithms in the value domain and the development of the temporal control structures (for the task activation schedule and the time-triggered communication) can be separated from each other.

Communication From Processing   Communication and processing are sequential activities based on different technologies. In a CPS the communication subsystem must transport data from the sender to one or more receivers within a given time constraint and with a required reliability without changing the data. A processing subsystem takes input-data and produces output-data using an application specific algorithm that must be executed on a given computer platform within a given time interval. If a CPS architecture guarantees the temporal separation of communication from processing and thus the independence of these two subsystems—as exemplified in a time-triggered architecture [Kop11, p.325]—then the complexity of the total Cyber-Physical System is significantly reduced. The evolution of the CPS with respect to changes in communication structure and processing technology can be simplified if some slack is provided in the respective time-triggered schedules.

Different Operational Modes   In a number of applications different operational modes can be distinguished that occur on different sections of the time line.

> **Example 7.2.5:** The control system onboard an airplane must handle four different modes of operation:
>
> - Taxiing on the airport
> - Take off
> - Cruising
> - Landing
>
> These modes of operations relate to different sections on the time-line. Each mode is nearly independent from other modes.

Batch Production Systems   In a batch processing manufacturing system, the work is divided into a number of sequential steps that are executed on different work stations. This organization of the work suggests a simple segmentation of the work in the temporal domain.

7.2.3   Abstraction

As outlined in Chap. 4 on modeling, an open system is characterized by an unlimited number of properties. An abstraction achieves the simplification by generating a model that deals only with those properties of the open system under discussion that are relevant for the given purpose and ignores all those properties that are not considered relevant for this purpose.

> **Example 7.2.6:** The abstraction of a *mass point* of a physical object ignores all properties that are related to the composition or appearance of the object (and many more). This abstraction of a mass point was crucial for the formulation of the laws of celestial mechanics.

Abstractions create language that allows people to focus on a particular aspect of a scenario and to communicate with other people without being detracted by irrelevant detail (which is not important in the context of the given purpose—*ceteris paribus*, see Example 2.3.4). It can be a challenging task to identify all those and only those properties of a given system that are responsible for a particular essential functionality (see Sect. 10.1).

Platform Abstraction   A platform consists of computing hardware, an operating system and, if necessary, middleware. It provides the environment for the execution of application software and abstracts from the hardware and the implementation of generic support services. The interface between the platform and an application software component is called *API (application programming interface)*. A platform can provide the following generic services:

- Run-time service for the execution of an application component
- Security services and authentication service

- Communication services
- Human-Machine Interface (HMI) service
- Data storage and data protection service including data recovery
- Error detection and fault-handling services, including reconfiguration

The API of a platform must be precisely specified and should be—in the face of system evolution—a stable interface.

A platform API for a Cyber-Physical Systems must contain the capability to specify the temporal parameters of the provided execution service, such that the given deadlines of the application can be met.

The generic platform services are useful for a multitude of different applications. The separation of the application software from the generic platform service simplifies the application software development.

Example 7.2.7: In the domain of smartphones the two platforms, Google's ANDROID and Apple's IOS dominate the market and provide the execution environment for millions of Apps.

Abstraction by Specification A specification of a part—e.g., a software component—is a contract between a user that uses the services of the part and the provider of the part. The user is satisfied with any part that meets the specification and abstracts from (is not interested in) the ways and means how the part provides its services. In real-time systems, the specification of a component must provide:

- A *requirements clause* (or preconditions) that informs about the necessary requirements on the input data,
- A *modifier clause* that informs about the changes to the state of the variables inside the component,
- An *effect clause* (or postcondition) that informs about the desired properties of the outputs and other effects of the execution of the procedure, and
- A *temporal clause* that informs about the physical time it takes to execute the component.

In software design, abstractions are introduced in order to separate the behavior of a program from its implementation and thus simplify the understanding of system behavior.

Abstraction of Models In some situations, a specific user may only be interested in those properties of a large model that are relevant for his purpose. He thus generates a simpler model (sometimes called a *view of a model*) that abstracts from all those properties of the large model (the target) that are not relevant for his purpose.

Example 7.2.8: A safety engineer must analyze the safety of behavior of a system under all possible use conditions. He is not interested in *comfort functions* that have no impact on the safety.

The topic of abstraction has been extensively discussed in Chaps. 4 and 5.

7.3   Static Structures

Static structures for the storage of data or the allocation of programs are developed at the time of design and remain constant during runtime. They are easier to analyze, test, and certify than dynamic structures that unfold during the execution of a program. Dynamic structures cannot be scrutinized until the concrete input data are known.

In the temporal domain, time-triggered events that occur periodically as a function of the progression of real-time provide a regular data-independent temporal control structure.

Example 7.3.1: The regular data-independent temporal control structure of the time-triggered architecture (TTA), combined with the use of anytime algorithms in the data domain, simplifies the reasoning about the temporal behavior of a TT-system.

The state space of dynamic structures that depend on the provided input data is orders of magnitude larger than the state space of static structures. According to the McCabe complexity metric (see Sect. 2.5) dynamic structure are thus much more complex than static structures. On the other hand, dynamic structures provide more flexibility. Dynamic structures should only be deployed if this added flexibility is absolutely essential in the given application context.

Example 7.3.2: A buffer overflow, caused by the erroneous allocation of dynamic data structures, is a common vulnerability that is used by intruders.

7.4   Emergence—New Conceptualization

The conceptualization of an emerging phenomenon (see Sect. 5.3) is the most powerful, but at the same time also the most difficult simplification principle. It requires the creation of *novel concepts* that capture only those properties of an emergent phenomenon that are of relevance for the given purpose and a mental disengagement from all those lower-level properties that are irrelevant at this higher level of abstraction. In contrast to *abstraction*, where existing concepts from the lower level that are considered relevant at the higher level are generalized and named by a new term, in the *conceptualization of an emergent phenomenon* the novel concept does not yet exist at the lower level (see also the important example 4.5.1).

Such a new conceptualization can transform a confusing scenario into a coherent and comprehensible one. This is an extremely creative mental act that requires intuition and a deep understanding of the application at hand, both from the point of view of technology and from the point of view of service provision in the given application context.

Example 7.4.1: Let us look at the new conceptualizations in models of the behavior of a computer [Avi82] considering the following four levels:

- The physical level
- The analog level
- The digital level
- The information level

At the physical level, we deal with the transistor effect in a semiconducting material. The transistor effect comes about by the meticulous distribution of different dopants in a semiconducting material that leads to a controllable electric field in a junction. The electrical characteristics, such as the *gain*, the *tolerated voltage* and the *frequency response* of a transistor depend on the kind and the precise distribution of the dopants in the semiconductor. At a higher level of abstraction that is concerned with the behavior of a transistor in an electronic circuit, we can consider a transistor as an electronic device that can amplify a small input signal that is applied between a pair of input terminals of a semiconducting material to a stronger output signal at its output terminals. We thus achieve a significant simplification by introducing the new concept of an amplifier to capture those properties of a transistor that are of relevance for the use of a transistor at the higher level of abstraction and leave behind all complicated considerations about the transport of electrons and holes in the semiconducting material.

At the analog level the electric signals of an electronic circuit are observed, such as the rise time of the voltages as the amplifiers perform their operations. The analysis of the circuit behavior at the analog level becomes difficult as soon as a multitude of amplifiers are involved in an electronic circuit—complexity develops. At this level the *new concept* of *logic variable* with the binary value of either True or False is introduced in a digital circuit. At this level it is also specified at what instants a circuit settles to a stable state, and what are the voltage bands that mean True or False in this stable state. Simplicity has been achieved by abstracting from the dense time and the analog values and by observing the circuit at the stable state only, such that the behavior of the circuit can be described by denoting the digital truth-values of the logic variables.

The next higher level, the digital logic level, introduces the new concept of a *Boolean operation* that operates on these logic variables. At the digital logic level, we describe the behavior of a logic circuit by the transformation of the state of logic variables by Boolean operations. A set of specified Boolean operations transforms the initial state and produces the final state—a sequence of bits at the end of the operation. The description of the relevant behavior of an electronic circuit is tremendously simplified.

Complexity creeps in again as we combine more and more logic circuits to build ever more complex digital electronic systems.

At the next higher level, the information level, we introduce the concept of a *data structure*, a structured sequence of bits that forms the data part of an Itom. Examples for data structures are pointers, real-valued numbers or pictures. Using the *procedure abstraction,* we introduce compound operations that operate on these data structures in order to realize the intended transformation: simplicity emerges again.

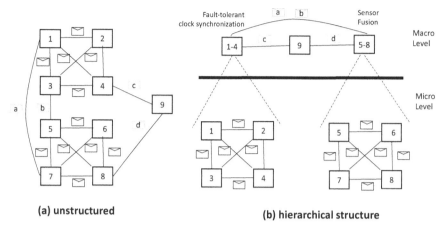

(a) unstructured

(b) hierarchical structure

Fig. 7.1 Example for a new conceptualization

The generation of a new concept that captures the essence of observed phenomena and puts aside all detail that are not relevant for the given purpose are at the core of scientific progress. The introduction of the concepts of *mass*, *acceleration* and *power* while disregarding the effects of *friction* forms the basics of Newtonian mechanics.

Example 7.4.2: In example 5.4.2 we have introduced the new concept of a fault-tolerant global time. At higher levels of abstraction we can use this new concept without regard for its implementation.

Example 7.4.2: In Fig. 7.1(a) the message exchanges between nine components of a distributed computer system are depicted in an unstructured organization. Figure 7.1(b) is a rearrangement of Fig. 7.1(a) to build a hierarchical structure with two newly conceptualized nodes: node (1–4) *fault-tolerant clock synchronization* and node (5–8) *sensor fusion*. These two nodes and node 9 are lifted to the macro-level, while the other nodes remain at the micro-level.

7.5   Points to Remember

- One fundamental problem in gaining an understanding of the behavior of a large artifact is the limited capacity of the cognitive apparatus of humans such that a *large monolithic pragmatic mental model (PMM)* that describes the full behavior of this artifact is not feasible.
- The most effective and widely deployed technique to simplify a large model is the *divide and conquer technique.*
- The critical issue in using the divide and conquer technique concerns the interactions among the simpler sub-solutions in the combined solution.

- We can divide a large model of a CPS along the following three orthogonal dimensions: (i) Partitioning in space (sometimes called *horizontal partitioning*); (ii) Segmentation in time (sometimes called *step-by-step analysis*) and (iii) Abstraction (sometimes called *vertical partitioning*).
- Essential functions should be implemented in separate (possibly hierarchic) components.
- In a time-triggered architecture, the development of the model algorithms in the value domain and the development of the temporal task activation schedules for the node-local operating system and the communication system can be separated from each other.
- A timely *edge-case detection* must be part of the normal operation; however, *the edge-case handling* should be contained in a separate subsystem that is just concerned with this edge case.
- A platform API for a Cyber-Physical System must contain the capability to specify the temporal parameters of the provided execution service, such that the given deadlines of the application can be met.
- The method of *step-by-step analysis* segments the behavior of a complex system into a sequence of successive steps on the timeline.
- In a safety-critical application we need at least three *independent subsystems,* the *normal processing subsystem,* the *safety assurance subsystem* and a *decision system* that decides whether the results produced by the normal processing subsystem are within a *safety envelope* that is established by the safety assurance subsystem.
- The software in the *decision subsystem that* checks whether the results of the *normal processing subsystem* are within the safety envelope provided by the *safety assurance subsystem* should be simple enough that it can be implemented with *simple software.*
- Static and regular structures that are fixed at design time and remain constant during the execution are easier to understand and certify than dynamic structures.
- The conceptualization of an emerging phenomenon is the most powerful, but at the same time also most difficult simplification principle.

Chapter 8
Communication Systems

"...with electronic communication, what one writes in a moment, eternity will not erase."

—Kent Alan Robinson

8.1 Introduction

Components in cyberspace interact by the exchange of messages that transport the data that must be shared by the components. The messages flow across communication channels according to the rules of the chosen communication protocol. The *payload* of a message, i.e., the user data in the message, is not modified during transport.

There are many different communication protocols in use in CPSs. The data can be coded by different means, e.g., electric, acoustic, and optical. These refinements of the transport mechanism are relevant when studying the implementation of the message transport mechanism at the physical level, but are irrelevant at a higher level where the only concerns are the reliable and timely arrival of the data in the message sent by the sender to the receiver(s).

8.2 Classification of Data

The choice of a communication protocol depends on the characteristics of the data that must be transported and on the physical parameters and constraints of the available communication channel. The most important distinction is between real-time data and non-real-time data. Recall that in cyberspace data is represented by a bit-pattern that informs, together with its explanation, about a property of an entity.

© Springer Nature Switzerland AG 2019
H. Kopetz, *Simplicity is Complex*, https://doi.org/10.1007/978-3-030-20411-2_8

8.2.1   Real-Time Data (RT Data)

RT data is used to control or monitor a physical process. The advance of a physical process depends on the progression of physical time. Therefore, the utility of RT data that is used in a cyber model that controls the physical process depends on the progression of physical time as well.

> **Example 8.2.1:** The information *the traffic light is green* that controls the flow of traffic across an intersection may only be used as long as the physical traffic light displays the green color.

The faster RT data is transported from its source to its destination the smaller is the reaction time between an observation of the physical world and the action that is based on this observation. Thus a faster RT data transport entails a shorter control cycle and a smaller model error (see Sect. 4.2.2). Since the total time required to transmit a message depends also on the length of the message (see Sect. 11.4 in the Annex) an effort must be made to reduce the length of a real-time message.

The distinction between essential data (e data) and context data (c data) that is part of an Itom (see Sect. 3.4) can help to reduce the length of data that must be transmitted in a real-time message. The e-data is dynamic and must be part of a real-time message, whereas the c-data is static and can be reduced by the proper alignment of contexts.

> **Example 8.2.2:** In an alarm message, the dynamic e-data is a single bit (alarm is on or off at the moment of observation). The static c-data must identify the source of the alarm. It is more efficient to identify an alarm (the c-data) by the position of the alarm bit in the alarm message than by an alarm name that must be carried in the alarm message.

In the following we elaborate on two different types of real-time data, *control data* and *alarm data*.

Control data is *periodic state data* that is used to control a physical process. Relevant state variables of the physical process are observed at equidistant periodic sampling points and delivered to a control algorithm in cyberspace that calculates the periodic set-points that are sent to the actuators at the end of the sampling interval (see Sect. 4.2).

The handling of *state data* at the endpoints of a communication channel is similar to the handling of a variable in a programming language. On reading, the present value of state data is copied and on writing the new value of state data replaces the old value.

Whenever fault-tolerance is realized by the replication of components, control data must be transmitted in a multi-cast topology to many receivers. It must be guaranteed that the failure of one receiver has no impact on another receiver or the sender of the message—neither in the value domain, nor in the temporal domain.

In many scenarios where real-time control data is transmitted *error detection* at the receiver is more appropriate than error detection at the sender.

> **Example 8.2.3:** Let us assume a configuration where a sensor node sends unidirectionally and periodically the observed real-time *sense data* to a central control room. In case the

communication channel between the sensor node and the control room computer is inter-
rupted, the control room computer—the receiver of the data— must detect the error. In this
scenario, error detection at the sender is useless, since the communication channel between
the sensor and the control room is broken.

The tradeoff between speed of transmission and reliability of transmission of
control data is delicate. We assume that the cyclic redundancy check (CRC) field in
the message is sufficient to guarantee with a very high probability that the contents
of the payload have not been mutilated during the transport. It thus follows that a
message arrives either correctly or not at all (*fail-silence* property of a message). An
error in the transmission of a *fail-silent message* can only be detected in the temporal domain. If the receiver knows exactly when a new version of a control message
must arrive, the receiver can detect the loss of a message immediately and the error
detection latency at the receiver is minimized.

A retransmission of a corrupted or lost message that contains control data makes
only sense if the time required for the retransmission is substantially smaller than
the sampling interval, after which a later and more relevant version of the control
data arrives.

Alarm data is *sporadic event data* (see Sect. 11.2) that informs about an unusual
or dangerous condition in the physical environment. An alarm may be silent for
years, but when it occurs it must be reported immediately.

From the perspective of the value domain, alarm data is simple: only a single bit
of data has to be transmitted. However, from the point of view of the temporal
domain, the transmission of alarm data is challenging, since the following temporal
requirements must be met:

1. The *worst-case transmission time of an alarm* message must be bounded,
   even in the case of an alarm shower where thousands of alarms are activated
   in a short temporal interval.
2. The *instant of occurrence of an alarm* must be reported with a good temporal
   accuracy, often in the μ-second range, since the temporal order of alarms is
   important from the point of view of causality analysis in an *alarm shower*.
3. The error detection latency of a fault in an alarm source must be known and
   bounded.

Remember that an *event* is a *change of state* (see Sects. 11.2 and 11.4 in the
Annex). A receiver can deduce the occurrence of an event by comparing successive
states. This is a well-established technique to guarantee the timeliness of single-bit
event data.

Example 8.2.4: The design of an interrupt system uses this technique. The voltage level of
the interrupt line is cyclically monitored (after the completion of every CPU instruction) in
order to detect a change of state, the *alarm event* signaled by the interrupt mechanism.

This technique of periodic state transmission and state monitoring by the receiver
can be used to handle alarm events in the communication system. Since the e-data
of an alarm is a single bit, the position of this bit in the periodic alarm message can

Table 8.1 Comparison of real-time data versus non-real-time data

Characteristic property	Real-time (RT) data	Non-real-time (NRT) data
Timeliness	Important	Not important
Preferred topology	Multicast	Point-to-point
Content of the payload of a message	State data	Event data
Error detection	At receiver	At sender
Loss of a single message	Not very relevant	Relevant
Retransmission of lost data	Not necessary	Necessary
Direction	Unidirectional	Bidirectional

be used to identify an alarm, i.e., the static c-data concerning this alarm. The example in Sect. 8.5 of this chapter demonstrates how alarm data can be transmitted with periodic state messages that meet the three listed requirements.

In conclusion, RT data transmission is time-critical and unidirectional, with error detection at the receiver of the message.

8.2.2 Non-Real-Time Data (NRT Data)

Non-real-time data (NRT data) comprises *file data* and *archival data*. Since no strict temporal constraints must be met in the transmission of NRT data, the tradeoff between speed of transmission and reliability of transmission is clear cut. The reliability of transmission and the integrity of the data have clear priority over the speed of transmission.

Normally NRT data is sent in a point-to-point message (or a sequence of point-to-point messages) from the sender to the receiver at an instant that is not known *a priori* to the receiver. Therefore, the correct receipt of the message must be relayed back from the receiver to the sender in an acknowledgment message, using a PAR (Positive Acknowledgement Retransmission) protocol. The sender monitors the timely arrival of this acknowledgement message by setting a timeout on the local clock of the sender to detect a communication failure.

In conclusion, NRT data transmission is not time-critical and bidirectional with error detection at the sender of the message.

8.3 Communication Protocols

The design of a communication protocol must consider the properties of the data that must be transmitted. The comparison of real-time data versus non-real-time data in Table 8.1 suggests that it will be difficult to design a unified protocol that meets the properties of both. An important differentiator between real-time data and non-real-time data relates to the handling of physical time.

8.3.1 Global Time

Independently of the requirements of a communication system, a distributed Cyber-Physical System (CPS) needs a *global time* of high precision for the following reasons:

- The modeling relation of Sect. 4.2 assumes that all sensors observe the state of the physical world at about the same instant of time. The same tick of a *global clock* should trigger the sampling actions that take place in the different nodes of a distributed computer system.
- The receiver of real-time data must ensure that the data is not used outside its temporal validity period. This requires a global time such that timestamps that are generated in one node can be interpreted by another node of a distributed system.
- The precise instant when an alarm occurs, e.g., the exact moment when a safety switch has tripped, is an important property of an alarm. In an analysis of an alarm shower that takes place in the central control room after many alarm messages have arrived, the precise instants of alarm occurrences help to find the causal event of an alarm shower. This requirement needs a global time of good accuracy to time-stamp an alarm locally in the node that observed the alarm.

One solution to the problems of global time is a central physical clock that provides the time signals to all nodes of a distributed CPS by dedicated time wires. However, such a central clock would be single point of failure (not considering the additional wiring effort).

An alternative solution is the provision of synchronized local clocks in the nodes of the distributed computer system. A fault-tolerant distributed clock synchronization algorithm achieves the periodic resynchronization of the local clocks. This solution avoids the *single point of failure problem* and the wiring problem but requires a communication system that generates the global time (see also Example 5.4.2).

Since global time must be available in a distributed CPS, this global time can be used to simplify the design of a real-time communication protocol.

8.3.2 Time-Triggered Protocols

The core idea of the time-triggered architecture and the time-triggered protocols is to use the progression of global time as the trigger for sending messages. The execution of a *time-triggered protocol (TTP)* provides a *periodic message transport service* for the transmission of state data. It can be characterized as follows:

(i) Sender and the receiver must have access to a global time of known precision.

(ii) Before the start of a time-triggered communication,

- A *periodic conflict free transmission schedule* must be supplied to the sender, the communication system, and the receiver. This transmission schedule contains the global times of the *periodic send instants* and the periodic *receive instants* of all TT messages.
- The sender of a message must provide to its communication interface the start address of the memory area from where the interface can fetch the payload data (state data) of the message.
- The receiver of the message must provide to its communication interface the start address of the memory area into which the interface can write the payload data (state data) of the message.

(iii) As soon as the global time reaches an *a priori* known send instant, the provided memory area of the sender is copied into the payload of a message by the communication interface and this message is sent to the receiver(s). The original data remains unmodified in the memory of the sender.

(iv) As soon as the message arrives at the receiver, the communication interface writes the payload of the message into the memory area that has been provided by the receiver, overwriting the previous contents of this memory area. This process is completed at the *a priori* known receive instant.

(v) Based on the knowledge about the expected receive instant of a message, the receiver must detect a transport error by not having received a new message at the *a priori* known receive instant.

Since no queueing effects occur in the transport of time-triggered messages and the preplanned schedule of the message in the transport system is guaranteed to be free of conflicts, the transport latency of a time-triggered message is small and known *a priori*. The error detection of TT messages is in the responsibility of the receiver, which simplifies the implementation of multi-cast communication required in a fault-tolerant system. A failure of a receiver has no impact on the operation of a sender. In conclusion, the services provided by a time-triggered protocol match ideally the temporal properties of real-time data.

The majority of the well-known communication protocols that are in use today (e.g., the internet protocols) have been designed for the transport of non-real-time file data. In the following Section we therefore focus on the less well-known time-triggered protocols for the transport of real-time data.

8.3.3 Time-Triggered Protocol (TTP)

TTP was the first time-triggered communication protocol developed at the Vienna University of Technology. In addition to the transport of real-time data, TTP implements a fault-tolerant distributed clock synchronization service and a membership service [Kop97].

During start-up TTP uses a central clock synchronization. It switches to a distributed fault-tolerant clock synchronization as soon as four nodes are available.

Distributed fault-tolerant clock synchronization is realized within TTP by the periodic execution of the following three steps:

1. Every node observes the *clock skews* between its local clock and all clocks of the other nodes.
2. Taking these observed time-differences as input, every node calculates a correction value for its local clock by executing a fault-tolerant synchronization algorithm.
3. Every node corrects the state of its local clock by the calculated correction value.

In TTP the clock skews are determined at each node by measuring locally—with the aid of hardware techniques—the interval between the *a priori* known expected arrival time of a message and the observed arrival time of a message. By making a clear distinction between the c-data derived from the time-triggered context and the e-data measured locally at each node (see Sect. 3.4), TTP does not require any extra communication bandwidth for the fault-tolerant clock synchronization service.

The TTP protocol is deployed in a number of safety-critical applications in the aerospace and railway domain. At the time of this writing, TTP has logged more than 10^9 hours of operating time in real-life environments.

8.3.4 Time-Triggered Ethernet (TTEthernet)

Ethernet is the most widely deployed communication protocol for non-real-time data. It is used for the transfer of small and large files in commercial and technical applications.

TTEthernet partitions the available bandwidth of an Ethernet system such that both RT-data and NRT-data can be communicated on the same physical channel. The switches will not start the transmission of a NRT message just before the *a priori* known instant when a RT message must be sent, such that the respective transmission links are idle and interference of NRT and RT data is completely avoided. The syntax of all TTEthernet messages is fully conformant with the established Ethernet standard.

TTEthernet chips that support a bandwidth of 1 Gbit/s on a twisted wire are available on the market. In such a network the transmission of an Ethernet message with a payload of 1000 bytes takes less than 10 μseconds.

8.4 Communication with the Controlled Object

Sensors and actuators form the interfaces between a component in cyberspace and a physical object in the physical world. They are controlled by an *interface component*.

8.4.1   Sensor Inputs

A sensor observes a property of a physical object, e.g., the temperature, and generates sense data (a bit pattern). The explanation of the meaning of this sense data is derived from the design and position of the sensor and the instant when the sense data was acquired. In order to simplify the further processing of the acquired data, its representation is transformed by the interface component to a standardized representation, the refined data, e.g. degrees *Kelvin* in the case of temperature.

The refined data can be sent to all components that need the data by periodic unidirectional multicast time-triggered messages. Since the receiver of these messages has provided to its communication interface the start addresses of the memory area into which the interface writes the arriving state data, the time-triggered communication system ensures autonomously that the latest version of the refined data is always available in the memory of the user of the information.

In order to detect the failure of sensor in a fault-tolerant system, at least three independent synchronized sensors are required to support a two-out-of-three voting schema.

8.4.2   Actuator

An actuator takes an Itom from cyberspace and produces the desired physical effect in the physical environment at the specified instant.

The proper operation of an actuator can only be determined if the desired physical effect, the stigmergic information (see Sect. 3.1) in the physical environment, is observed by one, or preferably more, independent sensors.

Example 8.4.1: A disregard of this design rule has contributed to the severity of the Three Miles Island Nuclear Accident [TMI18]. Within seconds of the shutdown, the pilot-operated relief valve (PORV) on the reactor cooling system opened, as it was supposed to. About 10 seconds later it should have closed. But it remained open, leaking vital reactor coolant water to the reactor coolant drain tank. The operators believed the relief valve had shut because instruments showed them that a "close" signal was sent to the valve. However, they did not have an instrument indicating the valve's actual position. The erroneous information about the state of the PORV valve, displayed in the control room, caused the operators to follow a mistaken mental model about the state of the physical process and to act according to this misinformation.

8.5   Real-Time Transfer of Alarm Data

The design of a real-time communication system for the timely transmission of sporadic alarms, even under peak load, is challenging. We take up the example of a power distribution network where a plurality of sensor nodes, located at remote

Fig. 8.1 Real-time
transfer of alarm data

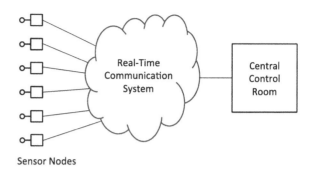

outstations, observe the environment and must send, in case of an *alarm,* an *alarm message* to a central control room within a guaranteed transport latency (see Fig. 8.1). We start with a discussion of the requirements that any implementation of such an alarm system must meet.

In this example we assume that all components of this system have access to a global time of adequate precision. In order to support fault-tolerance, the system in the central control room can consist of two or more replicated components.

The *state of a circuit breaker* is an example for an Itom that consists of the following atomic elements:

- The identification of the sensor that observed the alarm condition (c-data).
- The instant when the alarm occurred, expressed in a notation that uses the global time (e-data).

Any solution to this problem must meet the following requirements:

R1: The precise instant, when an alarm occurred must be reported to the central control room.

R2: An *a priori* known bound must bound the *reporting-latency,*i.e. the time interval between the occurrence of an alarm and the reporting of this alarm to the control room. This reporting-latency must also be guaranteed under peak load conditions, i.e. in case all sensor nodes send an alarm at about the same instant.

R3: The *error detection latency*, the time interval between a failure of a sensor node or the communication channel to the control room and the reporting of this failure to the control system must be bounded by an *a priori* known bound.

R4: The minimum time between the occurrences of subsequent alarms at a sensor node must be bounded by *a priori* known bound. If such a bound does not exist then this problem is unsolvable.

In order to support fault-tolerance, any solution to this problem must also meet the following constraints:

C1: An arbitrary failure of a sensor node must not interfere with the operation of the other sensor nodes or the components in the central control system.

C2: An arbitrary failure of a component in the central control room must not interfere with the operation of a sensor node or the other components in the central control.

If a shared real-time computer network is used to transport the Itoms between the sensor nodes and the central control room, then the satisfaction of the requirements R1 to R4, particularly requirement R2 about peak load performance, and the constraints C1 and C2 is not trivial. Since the use of the communication system is shared among many sensor nodes the transport latency in a shared real-time computer network cannot be neglected. Furthermore, an identification of the Itom and the precise global time of alarm occurrence must be part of an alarm message in addition to the single bit of the alarm.

Let us assume that in our power distribution network there are 100 out-stations within a radius of not more than 100 km from the central control room with 100 alarms in each out station, i.e., there is a total of 10,000 possible alarms. It is required that within a single cycle of the voltage, i.e. within 20 msec in a 50 Hz power grid, all activated alarms (which in peak load are all 10,000 alarms) must be reported to the central control room.

The available communication system is a time-triggered Ethernet with a bandwidth of 1 Gbit. The transport latency of a TTEthernet message over a distance of 100 km with a payload of 1000 bytes is the sum of the transmission time (about 10 μsec), the switching time in the time-triggered network (about 10 μsec), and the propagation delay (about 500 μsec). If a periodic multi-cast alarm-message, containing the precise instant of alarm occurrence of every alarm that occurred in an out-station, is sent with a period of 10 msec to the replicated components in the central control room, then the worst-case reporting latency BR is less than 11 msec. The precise instant of alarm occurrence—the e-data— that considers every (*a priori* known) instant of sending a time-triggered message—the c-data— as the start of *new epoch* can be coded in two bytes. The reported temporal resolution of an alarm event is then better than 1 μsec. The one hundred periodic alarm messages that are sent in each 10 msec period consume less than 5% of the available communication bandwidth of the TTEthernet. The remaining 95% of the bandwidth can be used for other real-time transmissions or for non-time critical file transfers via event-triggered file transfer protocols. A more detailed description of the transmission of alarm messages with a time-triggered communication system is contained in [Kop18a].

The time-triggered implementation of the alarm transport forms the basis of this splittable design. There is no temporal interference possible between the alarm system and the other system functions.

8.6 Points to Remember

• The choice of a communication protocol depends on the characteristics of the data that must be transported and the physical constraints of the available communication channel.

- Real-time control data must be transported from its source to its destination as fast as technically possible to reduce the reaction time between an observation of the physical world and the action that is based on this observation.
- The distinction between *essential data* and *context data* that is part of an Itom can help to reduce the length of data that must be transmitted in a real-time message.
- *Control data* is state data. *On reading*, the present value of state data is copied and on writing the new value of state data replaces the old value.
- When fault-tolerance is realized by replication, *control data* must be transmitted in a multi-cast topology to many receivers.
- For control data *error detection at the receiver* is more appropriate than *error detection at the sender.*
- A retransmission of a corrupted or lost message that contains *control data* makes only sense if the time required for the retransmission is substantially smaller than the sampling interval.
- *Alarm data* is *sporadic event data* that informs about an unusual or dangerous condition in the physical environment.
- For non-real-time data the reliability of transmission and the integrity of the data have clear priority over the speed of transmission.
- Independently of the requirements of a communication system, a distributed Cyber-Physical System needs a global time of high precision.
- Since global time must be available in a distributed Cyber-Physical System, this global time can be used to simplify the design of a real-time communication protocol.
- The services provided by a time-triggered protocol match ideally the temporal properties of real-time data.
- The Time-Triggered Protocol (TTP) does not require any extra bandwidth for a fault-tolerant clock synchronization service.
- Any interference between real-time communication and non-real time communication is avoided in TT Ethernet by design.
- An actuator takes an Itom from cyberspace and produces a desired *physical effect* in the physical environment. The proper operation of an actuator can only be determined if the desired physical effect is observed by an independent sensor.
- The precise global time of alarm occurrence must be part of an alarm message in addition to the *single bit of the alarm.*
- The time-triggered implementation of an alarm transport forms the basis of a splittable design, i.e. that the alarm system does not interfere with the other systems.

Chapter 9
Interface Design

> *"As far as the customer is concerned, the interface is the product."*
>
> —Jeff Raskin

9.1 Introduction

It is the purpose of an interface to provide access to the services of a system and to restrict the view into a system. The interfaces, placed in the boundaries—the skin (see Sect. 11.2)—of a system, should remain stable even in the face of system evolution. The proper design and placement of the interfaces ensures that minor changes in one part have no impact on the operation of other parts.

An interface of a CPS must enable the transport of Itoms across the skin of a system while keeping the semantic content of the Itoms invariant. The skin of a system provides a boundary between the internals of the system and the external world. Inside of the skin a uniform architectural style determines the representation of the data in the Itoms (the internal context). The representation of the Itoms outside the skin—in the external world— depends on the user of the system (the external context). Normally the substantial differences between the internal context and the external context necessitate a change in the representation of the data in the Itoms. These differences must be reconciled by an interface component.

In a Cyber-Physical System, we distinguish between three types of interfaces:

- *Human-machine interfaces*: a human uses the services of a computer or controls a machine.
- *Machine-to-machine interfaces*: one computer may use the services of another computer or both computers are on an equal footing.
- *Process-control interface*: a computer *observes* or *controls* a process in the physical world.

An interface can be studied at different levels, where the level determines what is the intent and object of the study.

© Springer Nature Switzerland AG 2019
H. Kopetz, *Simplicity is Complex*, https://doi.org/10.1007/978-3-030-20411-2_9

Example 9.1.1: A Human-machine interface (HMI) can be studied at the level of the physical interactions between a human and a computer (e.g., visual, sound, haptic) or at the level of information representation or at the level of information content, i.e., the semantics of the Itoms.

The focus of this chapter is on the semantic level.

9.2   Architectural Style

The *architectural style* refers to all explicit and implicit principles, rules and conventions that are followed when designing the inside of a system e.g., the representation of data, protocols, syntax, naming, and semantics of variables. In order to achieve a good context alignment between all components inside a large system and all members of a large project team, a precise documentation of the architectural style should be available at the start of the design process.

A thorough context alignment inside a system reduces the amount of c-data (see Sect. 3.4 on the difference between c-data and e-data) that must be transmitted among the components of a large system.

The following topics should be dealt with in the architectural style document:

- *Standard format for the representation of data*: It is good practice to adhere to international standards, e.g., ISO standards for the representation of refined data, of time and other physical qualities. The representation of sense data is determined by the design of a sensor/actuator and should be hidden inside an interface component.
- *Naming convention for variables:* The name of a variable establishes a link between the data in the variable and the explanation of the data in the conceptual landscape of a user or the software that uses the data. Utmost care should be taken to assign systematically meaningful names to variables
- *Communication protocols*: A standard protocol that supports the real-time communication among components and non-real time communication for file transfer should be specified. It is good practice to adhere to widely used standard protocols, such as Ethernet and its real-time variant TTEthernet, since every new protocol increases the accidental complexity.
- *Error detection strategy*: Since every computational action or every communication action can fail, mechanisms must be in place to detect component and communication failures. In a Cyber-Physical System, a failure occurs if an expected result is not provided before the deadline or if there is an error in the value domain of the result. A failure can only be detected if the calculated result can be compared with a reference. It must be stated how this reference for error detection is established. Error detection should have a low latency in order to be able to start a mitigating action before the error has propagated to other parts of the system.

- *Error handling strategy*: Error detection and error handling are different activities that require a separation of concern. Transient hardware errors and *Heisenbugs* in the software (see Sect. 11.4) have the same appearance and require the same error detection and short-term error handling mechanisms, but different error elimination strategies.
- *Testing strategy*: The proper operation of every component must be testable in a stand-alone context and in the context of the integrated system. This requires multi-cast communication protocols that support the observation of the inputs and outputs of every component during operation without a probe effect. Emergent phenomena, particularly those caused by a stigmergic information flow, cannot be detected during the stand-alone tests of the components. It is thus of paramount importance that the project schedule provides enough time for the integrated system tests in a real-life environment.

9.3 Interface Component

Figure 9.1 depicts a CPS that contains two internal components, *component A* and *component B*. Component A communicates across interface component A to a human user. Component B communicates across interface component B to the physical process.

The system-internal interface between component A and component B is realized by the *internal interface component* in the middle of Fig. 9.1. If both internal components (component A and component B) adhere to the same architectural style then the internal interface component is simplified.

The external view of the interface component A is determined by the conceptual landscape in the mind of the human user. The interface component A transforms the interface data that conform to the external view of the human-machine interface (HMI) to the representation prescribed by the given architectural style (internal view) and vice versa, while keeping the semantic content of the data invariant. Section 9.5 elaborates on the design of the external view of an HMI.

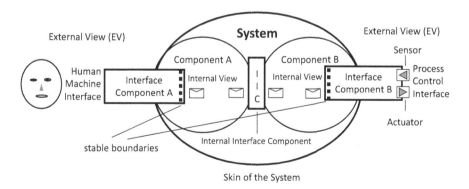

Fig. 9.1 Skin of a system with interface components

The external view of the interface component B is the world of physical quantities at the process control interface. Sensors convert these physical quantities, the
physical sensations (see Sect. 3.2), to bit-patterns (sense data) that are transformed
by the interface component to a representation that conforms to the prescribed
architectural style (refined data) of the system. In the opposite direction the actuators provide the output to the physical object of the CPS.

In a distributed CPS the internal interfaces are *message based*. An interface
should remain stable, even in the case of system evolution.

Example 9.3.1: Consider a request to change the language of the HMI in Fig. 9.1. Such a
change should be confined to the external view of the interface component and should not
have any impact on the internal view of the interface component A.

If periodic *time-triggered state messages* provide the data transport across the
interfaces inside a system then the *temporal stability* of the internal boundaries is
guaranteed by design.

9.4   Interface Types

A component can support a number of different interfaces, where each one of these
interfaces serves a different purpose. By cleanly separating the interfaces according
to purpose, we can simplify the design of each one of these separate interfaces—an
application of the simplification technique *divide-and-conquer.*

9.4.1   Service Interface

The most important interfaces of a component are its *service interfaces,* i.e., the
interfaces where the component grants access to the services it is offering. A service
is an identified *unit of functionality* of a component that is provided via a specified
message exchange between a *service user* and a *service provider.* In some cases,
service user and service provider enter into a *service contract* that establishes the
technical and commercial conditions of the service. The specification of the functionality of the service interface should be ignorant about the implementation of the
service. It must limit the access privileges of the service user to those resources that
are specified in the service contract.

Example 9.4.1: The difference between the terms *function* and *functionality* is important
and can be explained by looking at a component that is intended to calculate the *income tax.*
The *functionality* of the component refers to the calculation that the component is intended
to perform—*correctly calculate the current income tax.* The *function* refers to the calculation that the component actually performs. If the tax code changes, the functionality remains
the same, but the function must change.

In Cyber-Physical Systems, a service interface can often be implemented as a *unidirectional information flow interface*, where—at the semantic level—the service provider functions independently of the service users and presents to the service user a stream of time-triggered messages containing state data. To reduce the coupling among subsystems, such a semantic unidirectionality should be maintained throughout all levels of an implementation.

> **Example 9.4.2:** A clock synchronization service can be provided across a unidirectional service interface. The *time service provider* broadcasts periodically a message that contains the current value of the global time.

Periodic time-triggered state messages provide the simplest form of a service interface for the unidirectional transport of periodic information. Wherever possible, this form of a service interface should be deployed in order to maintain the independence of the service provider.

> **Example 9.4.3:** A sensor component that observes a physical quantity can provide the refined data to other components that use this service by a stream of unidirectional periodic multicast time-triggered messages. The service user will always find the most recent value of this quantity in its memory without any additional synchronization requirements.

Unidirectional periodic state message can also be deployed to present the state of a system at the *Human Machine Interface (HMI)* that monitors the operation of a CPS. The *HMI component* that controls the HMI can contain a real-time data base of state variables that is autonomously and periodically updated by the time-triggered messages from the interface components that observe the state of the controlled object and from the control components that provide periodically the set points sent to the actuators.

9.4.2   Configuration and Control Interface

The *configuration and control interface* is required to arrange the configuration of a component, to set appropriate execution parameters, such as the *power level*, and to start or terminate the execution of a component. The configuration and control interface is part of the operating system that provides the execution environment of a component. The user of the configuration and control interface does not have to know the internal implementation of a component.

9.4.3   Maintenance Interface

The maintenance interface opens a window into the internal implementation of a component. Only a trained maintenance engineer who knows the internals of the component implementation is capable to work across the maintenance interface to correct errors in the design of a component or to modify the functions of a component. Access to the maintenance interface should be tightly controlled and precisely documented.

9.5 Interface Model

As soon as the purpose of an interface has been precisely defined, an *interface model* must be developed. The interface model is—as all models (see Chap. 4)—an abstraction that contains and explains those–and only those– properties and functionalities of the component that are required to use the interface by an interface user.

For a user, the interface model is the system Any change in the functionality of an interface model has severe consequences for all parties that use this interface.

The human machine interface (HMI) model provides to the user a window into a system. It is important to restrict the view through this window to only those entities within the system that are required to serve the intended purpose of the interface and make other entities invisible. Otherwise, a user may use these other entities and restrict the freedom of the designer to change the implementation in case of required evolutionary changes.

In machine-to-machine communication the interface model should be formal such that the correctness of the interplay between both interfacing machines can be formally analyzed and interpretation errors by humans are eliminated.

9.6 Human Machine Interface (HMI)

Every Human-Machine Interface (HMI) of a Cyber-Physical System (CPS) is developed to support the interactions between the CPS and a human user who wants to achieve a given purpose. The results of the purpose analysis are the essential and supplemental functionalities that must be supported by the planned HMI. Each functionality requires the conception of a corresponding mental interface model (MIM, see also Sect. 4.3 on mental models) that must be acquired by an operator in order to use this HMI function correctly. If there is a *multitude of functions* supported at an HMI, then a *multitude of MIMs* is required. Although each one of these MIMs should be independent (divide and conquer), they all should form a consistent whole governed by the same architectural style, i.e., appearance and style of interaction.

The design of each MIM must take account of the finite capacity of the human mind and the limited reaction time of humans. Every MIM of the multitude of MIMs should be self-contained (as far as possible) and must include the relevant Cyber-Physical System (*CPS*) states and the relevant states of the environment that an interface user must know in order to correctly assess the current situation of the system in its present environment. It must also specify the state transitions from the present state to future states, pointing out which of these state transitions are in the sphere of control of the user. In order to provide full information to the user, the present state of the targeted subsystem of the CPS and its associated environment should be visible at the HMI at all times.

In case of a large system it may be necessary to develop a multi-level hierarchy of MIMs (see Sect. 5.2). Every one of the mental models in a multi-level hierarchy should be comprehensive—it must include a display of all Itoms that are of relevance at this level and all operator actions that can be carried out at this level. In order to eliminate horizontal dependencies, the multi-level hierarchy should be a formal hierarchy.

An operator who acts at a *Human-Machine Interface (HMI) of a CPS* and looks, from his vantage point, at the behavior of the CPS in its environment must align the *mental interface model* (MIM) in his mind with the working of the CPS. This mental interface model receives two types of inputs, stigmergic inputs and cyber inputs.

The stigmergic inputs consist of the relevant Itoms about the state of the environment identified by the human sense organs without any support from the cyber systems.

Example 9.6.1: For example, a driver of a car looks at the environment, sees the other cars on the road, some of them blinking, hears a horn blowing, and feels a bump on the road—We call all of these inputs stigmergic inputs.

The cyber inputs consist of the Itoms presented to the operator at the HMI by the cyber system.

Example 9.6.2: An example of a cyber input to the driver of a car is an icon of car, displayed at the dash-port of the car that informs the driver about a car driving at a certain distance ahead even if, due to fog, the direct human vision cannot identify the object in front by the corresponding stigmergic input.

The outputs of the mental interface model are normally—but not always—directed to the HMI of the CPS. They are processed and checked by the cyber system before they are released to the actuators that control the physical process.

A well-designed HMI must be forgivable, since humans make errors quite frequently. A forgivable HMI informs the user about the consequences of a requested action and provides the capability to go back in case of an error. It issues a warning in case the user requests an instruction that cannot be undone. A forgivable HMI gives a user the capability to directly return to a *defined initial state* from any other state with a simple command or after a pre-determined time-out.

Some mental effort is required to learn about the structure and behavior of the mental interface model. Two methods of learning can be distinguished: (i) *studying of the operating manual* and (ii) *trial and error* by practical experience. Many individuals—including myself—prefer the trial-and-error method to the studying of the operating manual. The designer of an HMI should take note of this fact and provide, for a novice user, a special learning mode of operation where operator errors are immediately detected and the novice user is guided to the correct operation of the system by explaining why the past action was erroneous.

As soon as a correct mental interface model has been formed in the mind of a user, the experienced user will not appreciate that he has to go to a tedious learning mode every time he is using the system. For the experienced user an effective procedure to control the relevant parameters of the CPS should be provided.

Normally there is a third user group at an HMI, the *expert users*, who have a deep understanding of the internals of the CPS and can set internal parameters that determine the *safe envelope of operation* of the system. In safety-critical systems, the actions of the expert users are thus *safety relevant*. These expert users must go through a documented authentication procedure to acquire the privilege to change these internal parameters in the expert mode. Even an experienced user should not be allowed to enter the expert mode without a proper documented authentication.

In order to give a user the illusion of an immediate action, the response time of a safety-critical system between a control input at the HMI and the start of the corresponding action in the physical world should be in the order of 100 msec. This is considerable shorter than the average response time of 500 msec required from Web applications. The neglect of an adequate response time has led to serious accidents.

Example 9.6.3: The accident report about a crash of a Gripen airplane [Gai89] states: "… the pilot control commands where subjected to such a delay that he was out of phase with the aircraft's motion. This caused uncontrollable pitch oscillations, which resulted in the aircraft hitting the runway in the final phase of the landing."

9.6.1   Alignment of Contexts

The most important means for the simplification of a human-machine interface is *context alignment* between the purpose and the context of the user with the representation of the Itoms at the human-machine interface by the machine. The cultural background, the personal experience and the given circumstances of the user in his current environment form the conceptual landscape of a user. If the machine is aware of this conceptualization of the user and adapts the representation of the Itoms at the HMI accordingly, then the comprehension of the provided Itoms by the user is substantially facilitated.

According to the reciprocity principle discussed in Sect. 3.4, the proper alignment of contexts reduces the amount of data that must be exchanged in a conversation in order to convey the relevant information. In case full alignment of contexts is achieved, only the essential data (e-data) has to be exchanged across an interface. The smaller the amount of data that must be exchanged across an interface to achieve the desired purpose, the simpler it is to use the interface.

Cultural Differences

Many Cyber-Physical Systems (CPSs) are designed for the global worldwide market, where operators that speak different languages and have disparate cultural backgrounds must learn to use a machine. The cross-cultural differences in the representation of Itoms (e.g., familiar icons, common measurement units, color-related emotions, data formats), the structure of language (characters, font size, direction of writing, naming) and common dialog characteristics (navigation concept, interaction-path, communication speed) are substantial [Hei07]. Furthermore, there are contrasting value systems (e.g., personal responsibility. Risk avoidance, face saving in case of an error) in different cultures [Hof17]. If the CPS is aware of these cultural idiosyncrasies of the user's environment and presents the data of an Itom in a form that the user feels at home in front of the given machine, then the mental effort to comprehend the information and the probability to make operator errors is substantially reduced—i.e. simplification is achieved by the software behind the human-machine interface (HMI).

> **Example 9.6.4:** The data in an Itom that denotes the energy content of gasoline can be presented in BTU per gallon, in kilocalories per pound or in kWh per liter. The cultural background of a user determines which one of these presentations is most understandable to an attendant user.

Situational Awareness

The alignment between the context of a user and the context considered by the CPS can be further improved if the CPS is aware of the current situation. Time and position derived from a navigation system, such as GPS, and possible other sensors (e.g., cameras, microphones) inform the CPS about the current situation. Combined with an immediate access to situational information stored in the cloud, the CPS becomes aware of the current environment of the user and can use this knowledge to present relevant Itoms and alternatives for action appropriately at the HMI.

A multitude of IoT (Internet of Things) sensors in a CPS, coupled with Artificial Intelligence technology can help to figure out the current context of the user in a communication scenario and thus support the context alignment.

Personal Profile

In applications, where the authenticity of the partner must be established before a transaction can take place, (e.g. in a banking application, or at a medical doctor) the personal data and the stored user profile of the partner can be utilized to further align the context between the authenticated partner and the CPS.

Standardization

A standardized general schema can reduce the complexity of orientation in a new environment. This schema can be filled in with the appropriate details that characterize the concrete situation. Since the general schemas are likely to be already in the conceptual landscape of a user, the adaptation to a particular context can be achieved by the simple mental process of assimilation and does not require the more difficult mental process of accommodation (see Sect. 2.4).

Example 9.6.5: The international sign for a power switch tells a user independently of her cultural and language background the meaning of the switch in the given context. The sign, which is actually an icon, provides a pictorial explanation for the use (the data) in the concrete situation.

Example 9.6.6: There are actually two types of power switches in use: an event-based power switch (a push button) and a state-based power switch (a toggle switch with two discernible states). The meaning of the push button depends on the state of the involved device. If the power is on, the meaning is: *turn the power off*. If the power is off, the meaning is: *turn the power on*. The meaning of a toggle switch is: one state for the power is on, the other state for the power is off. From the point of view of cognitive complexity, the toggle switch is significantly simpler, because its meaning is self-contained.

In 1928 Otto Neurath of the Vienna Circle designed a visual language for public communication consisting of a set of icons (pictograms) that provide a means of explanation that transcends the language barrier. The International Standards Organization (ISO) has taken up parts of this effort in the *international language of ISO graphical symbols* [ISO13], where more than 4000 pictograms characterizing a variety of situations have been standardized.

In many disciplines international standards are available that specify the representation of physical quantities. If all communicating subsystems adhere to an agreed-upon set of such standards, then the representational incompatibilities at an interface disappear. However, it is not only the representation of items that must be standardized. Also, the exact meaning of the *words* that are used in the design of an application must be aligned. The topic of worldwide standardization is getting more important, as systems from all over the globe interact via the Internet.

9.7 Points to Remember

- It is the purpose of an interface to provide access to the services of a system and to restrict the view into a system. The interface model is an abstraction that contains and explains those–and only those–*properties and functionalities* of a component that are required to satisfy the established purpose of the interface.
- In order to achieve a *good context alignment* between all components inside a large system and all members of a large project team, a precise documentation of the *architectural style* should be available at the start of the design process.
- By cleanly separating the interfaces of a component according to function, we can simplify the design of each one of these interfaces (*divide and conquer*).
- The most important interfaces of a component are its *service interfaces, i.e.,* the interfaces where the component grants access to the *functionality* it is offering. *Function* is different from *functionality*.
- Any change in the functionality of an interface model has severe consequences for all parties that use this interface. Therefore, the functionality of an interface should remain stable—as far as technically possible— even in the context of system evolution.

- In Cyber-Physical Systems, a service interface can often be implemented as a *unidirectional information flow interface*, where—at the semantic level—the service provider functions independently of the service users and presents to the service user a stream of time-triggered state messages
- Each one of a multitude of mental-interface models (MIMs) should be independent (divide and conquer), but they all should form a *consistent* whole governed by the same *appearance* and *style of interaction*.
- Many individuals prefer the *trial-and-error* method to the *studying of the operating manual* when confronted with a new interface.
- The most important means for the simplification of a human-machine interface is *context alignment* between *the purpose and the context of the user* with the *representation of the Itoms at the human-machine interface by the machine*.
- The cross-cultural differences in the pictorial representation of Itoms – the structure of language and common dialog characteristics – are substantial and must be considered in the design of a *mental interface model* for a global product.

Chapter 10
Simplification of a Design

> *"You know you have achieved perfection in design, not when you have nothing more to add, but when you have nothing more to take away."*
>
> — Antoine de Saint-Exupery

10.1 Introduction

The final chapter of this book elaborates on how to apply the understanding gained by the study of the previous chapters to design a simple Cyber-Physical System. The personal experience of the author in the design and management of Cyber-Physical System projects in industry and academia shaped to a considerable degree the insights discussed in this chapter.

As already remarked in the beginning, it requires deep technical insight and a strong character to build a simple system. During the first phases of a new design project, the conceptual integrity of the evolving design must often be defended against a multitude of nice-to-have features that are of marginal utility for the essential purpose of the system. These supplemental features are likely to lead to unanticipated feature interactions and corrupt the original conceptual integrity envisioned by the designer. This is one reason why we consider a thorough purpose analysis and a good dose of requirement skepticism as the corner stones for the design of a simple system.

10.2 Purpose Analysis

The *NASA Study on Flight Software Complexity* [Dvo09] makes a distinction between *essential functionality* of a system which comes from vetted requirements and is therefore unavoidable, and *incidental complexity*, which arises from misguided decisions about architecture, decomposition, and implementation and which can be reduced by making wise decisions during the design and development process of a system. This is similar to the distinctions Fred Brooks makes in his paper

© Springer Nature Switzerland AG 2019
H. Kopetz, *Simplicity is Complex*, https://doi.org/10.1007/978-3-030-20411-2_10

No Silver Bullet [Bro87] between *essential complexity* and *accidental complexity*. Essential complexity, which is related to the essential functionality of a planned system, is inherent and unavoidable, given the purpose of the intended system, while accidental complexity is the consequence of unwise design decisions.

Every Cyber-Physical System is conceived to serve a given purpose. The thorough analysis and the precise documentation of the purpose of a system must be at the start of every project, since the purpose of a system determines the essential functionality and, in consequence, the essential complexity. The purpose of a system defines the telos, the goal that is required for a teleological explanation (see Sect. 2.3) of the system functions.

The identified essential functionality of an intended system should be examined as a service in a chain of services required to realize the end-to-end experience of a user. Such an endeavor forces the designer to place the functions of the intended system in a wider context. It can lead toward the discovery of new solution alternatives.

> **Example 10.2.1:** Consider the example of the design of a new ticket vending machine for the public transportation system in a city. Customers have complained over the long queues before the vending machines. There are a number of options to reduce the time spent for buying a ticket: increase the number of vending machines, install a faster ticket printer, or simplify the dialogue by using stored information about the travel history of a known traveler (identified by his credit card). A field experience study has shown that it is mainly the tourists that take a long time to orient themselves before a vending machine that is new to them. Selling a multi-day ticket at the airport that allows a tourist to travel freely in the city would solve the problem by a completely alternate approach.

In many projects, the given *initial requirements specification*—which are often part of a commercial contract—are taken as an unchallenged reference document for the determination of the essential functionality. This is a doomed starting point for a project.

It is the purpose of the system and its goals, derived from the purpose analysis, that determine the essential functionality.

The hasty acceptance of given requirements without any questioning of the rationale behind them and their relations to the essential purpose of the user of the system introduces a down-stream essential complexity that cannot be mitigated any more in the following design process.

In many cases different stakeholders have differing views about the purpose of a planned system leading to conflicts at the level of system goals. These conflicts must be resolved at the level of the purpose analysis, which establishes the measurable goals of the planned system. Since the purpose analysis is at a higher level of abstraction than the requirements analysis, it is easier to resolve developing conflicts among the stakeholders at the level of the purpose analysis.

At the end of the purpose analysis the system goals should be prioritized. This can be achieved by a sequence of *pairwise comparative judgments* of the identified goals by all stakeholder. A detailed purpose analysis that provides a prioritized list of system goals that is accepted by all stakeholders is a good starting point for a

large project. It provides a consistent documentation of the appraised essential functionality that is expected from the intended system without going into the details of
design decisions. The documentation about the prioritized list of system goals
establishes the rationale for the development of the later system requirements.

In every large system project there will be a multitude of different requirements.
Based on the prioritized list of system goals, these requirements should be classified
at least into two groups, *essential requirements* and *supplemental requirements*. The
essential requirements relate to essential system functions. Only the satisfaction of
the essential system functions is indispensable for the success of a system.

Every requirement should be accompanied by a precise acceptance test. If it is
not possible to specify a precise acceptance test that can determine at the end of the
project whether the requirement has been satisfied or not, then this requirement is
meaningless and should be discarded.

At the start of a project it is often unknown which requirements are easy to
implement and which requirements cause a substantial implementation effort. As
more information becomes available about the effort needed to implement a given
system function, an *iteration* concerning the requirements specification must take
place. Supplemental requirements, which demand a substantial implementation
effort, should be discarded. But also, the essential requirements that are difficult to
implement should be questioned in order to find out if their relation to the primary
system goals justifies the envisioned large implementation effort. After a number of
requirements iterations that change some of the original requirements, and after
considering the result of trade-off studies about the efforts required to solve an identified problem by hardware or by software, the *essential hardware requirements* and
the *essential software requirements* of a project can be established.

At this time the essential complexity of a system should have been reduced to its
essence.

10.3   System Evolution

A large system evolves gradually out of a smaller system that has been deployed
successfully. The operation of a successful system changes its environment in unexpected ways. These unforeseen changes lead not only to new requirements that must
be considered in the next version of the system, but also to the elimination of some
of the previous requirements. This step-by-step evolution of large systems can be
observed not only in the domain of computer technology (e.g., the evolution of the
Internet) but also in many other domains, e.g., in biology or in society [Mat92].

Experience has shown that the future evolution of a successful system is inevi-
table. It is thus prudent to plan for this evolution already during the architecture
design phase. There are three main causes for system evolution: (i) changes of the
environment of a system, (ii) changes of user requirement and (iii) changes of the
implementation technology. The scope and probability of such changes should be
estimated before the architecture design.

We call a change request that impacts only the internals of a component but has no effect on the interfaces of a component a *minor change*. A change request that requires the modification of a component interface is a *major change*. Major change requests entail modifications of all those components that use the affected interfaces.

The proper placement of component interfaces during the architecture design phase can help to reduce the later need for major changes. Major changes, if not accompanied by a careful change management process, can lead to an erosion of the architecture [Fur19].

10.4 Architecture Design

The architecture design phase begins with the conceptualization of the *top-level structure* of the planned system.

10.4.1 Top-Level System Structure

In Fig. 10.1 the *top-level system structure* phase of a drive-by-wire application is depicted in the form of a unidirectional data flow diagram.

In this application the environment is periodically sampled by the *data acquisition components*, the sampled data are fused by the sensor *fusion component*, the planned trajectory is determined by the *trajectory calculation component* and the outputs are delivered to the actuators by the *output component*.

In case it has been decided that a time-triggered architecture is to be implemented, a first version of the temporal control structure (TCS) of this application should be developed in the next step. The TCS specifies the instants when a component is activated and when a time-triggered message must be sent.

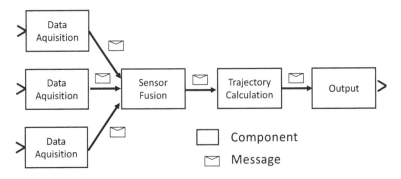

Fig. 10.1 Unidirectional data flow diagram

In order to develop this TCS, a first estimate of the following parameters is needed

- The *bandwidth* of the communication system.
- The *duration of the sampling intervals.*
- The *length of the messages* that must be exchanged along the communication channels.
- The *service intervals SI* required for the execution of the software components on the chosen hardware platform.

The TCS is calculated by a global scheduler and provides *temporal constraints* for the implementation of the components. Preferably, the components should deploy anytime algorithms to produce satisfycing results by the given deadline. The TCS partitions a large system into a set of nearly independent components and ensures that the system integration of the components will not be hampered by temporal side effects.

Example 10.4.1 A time-triggered (TT) computing platform can hide the real-time communication behind the *platform abstraction* (see Sect. 7.2). Based on the TCS, a *TT-platform* can write the input data in a specified memory area of an application component just before the time-triggered execution of this application software component is started and can fetch the output data from a specified memory area of the application component as soon as the Service Interval (SI) has ended. The application software is simplified because it does not have to execute Input/output commands. It will always find at the right instant the *valid real-time input data* in its memory.

As soon as more realistic estimates of the temporal parameters are available, an enhanced version of the TCS can be produced. Whereas the first version of the TCS will be based on a rough guess of the required bandwidth based on a first estimate of the size of Itoms that must be transported, the detailed design will look at relation of the e-data and c-data (see Sect. 3.4) within an Itom and will try to minimize the length of bit-pattern that must be transported in order to optimize the responsiveness.

The following further issues must be considered during the architecture design phase:

10.4.2 Independent Components

If at all possible, every essential requirement should be assigned to an independent top-level component that implements only this essential requirement and nothing else [Suh90]—see Sect. 7.2. This assignment of the essential requirements to independent components with carefully controlled interfaces among these components simplifies the later evolution of the system functions.

10.4.3   Scheduling Slack

This issue is relevant in the design of a time-triggered system only. We call the uncommitted time intervals in the temporal control structure (the TCS) of a time-triggered system the *scheduling slacks*. The scheduling slack in the service interval SI (see Sect. 11.3) makes it possible to accommodate modifications of the respective computational actions without any change of the communication schedules. The scheduling slack of a communication action allows a path change of the message route without an effect on the nodes that execute the control algorithms.

10.4.4   Peak Load Behavior

Peak-load behavior refers to the behavior of a system given all possible stimuli (e.g., alarms) occur at the same moment. In many applications the predictable services of a control system are required most urgently when an abnormal situation develops, causing an alarm shower. The proper reaction of the system to such a situation, including the filtering of the alarms such that only the most important alarms are immediately presented to the operator, is absolutely essential. There are many cyber-physical control systems in operation, where the peak-load performance is unpredictable—the probability of failure of such a system is highest, when the services of the system are needed most urgently. A good design rule says: *if the system behavior is predictable under peak-load conditions, then the normal load will take of itself.*

10.4.5   Fault Management

Eventually, any hardware unit will fail—either in a transient or permanent mode—and any large software system will contain undetected design errors in the software. Considerations about the proper detection, handling and mitigation of these faults are of paramount importance during the design of the system architecture. The first consideration deals with prompt error detection. Prompt error detection makes it possible to immediately mitigate the consequences of an original fault before error propagation has infected other healthy system components. If a component exhibits a *fail-silent behavior*—either the component operates correctly or does not produce any result at all—then it is possible to realize error detection by checks in the temporal domain without the need to look at the value domain. The *a priori* known schedule of a time-triggered architecture simplifies the detection of the failure of a *fail-silent component* in the temporal domain.

10.5   Multi-Level Hierarchy

In Chap. 5 on *Multi-level Hierarchies* we discussed extensively the characteristics of a multi-level hierarchy and conjectured that the construction of a multi-level hierarchy of models is one (maybe the only one) effective means to get a handle on the complexity of a large system. Each level model must be small enough that a human can comprehend it.

The title of this book *Simplicity is Complex* will be fully appreciated as soon as the designer becomes aware of the enormous mental effort required for this design step, the construction of a multi-level hierarchy of models to arrive at an understandable structure of a large system. Since there is a danger that this effort leads to *paralysis by analysis,* the resources allocated to this design step should be limited beforehand. Even a cursory attempt to develop such a multi-level framework will be worth the effort, since it widens the view on the system and its place in its overall environment and the chain of services.

Fig. 10.2 is the result of an effort to find a multi-level hierarchic structure for the overall transport system, consisting of the road-traffic system, the rail-traffic system, the air-traffic system and the water-traffic system. Only one facet has been chosen for a detailed decomposition: the position of the automobile in the overall transport system (Fig. 10.3).

Example 10.5.1: Fig. 10.2 depicts an eight-level hierarchy of models that establishes the framework of a multi-level hierarchy for structuring the worldwide transport system. The focus of Fig. 10.3 is on the decomposition of an automotive system. At *level n-2* we decompose the road-traffic system into road traffic participants, the road system and the traffic code. A road traffic participant has the following relevant properties

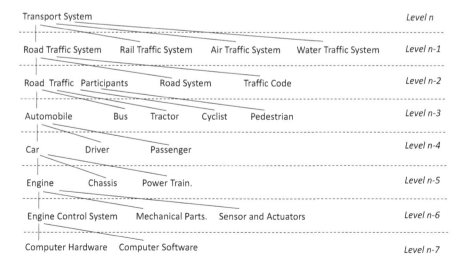

Fig. 10.2   Multilevel hierarchy of models of the transport system

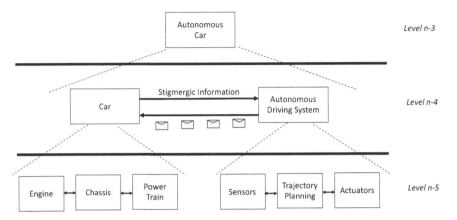

Fig. 10.3 Multilevel structure of an autonomous car

1. Needs exclusive free space for its volume to move
2. Has a mass and the force to move the mass
3. Has the capability to control the movement
4. Has a sensor system to observe the environment
5. Has knowledge of the road system
6. Has a plan to reach a goal
7. Has a goal.

These properties are characteristic for all road traffic participants (even for a pedestrian). In the case of an automobile these seven properties of a road traffic participant are allocated at *level n-4* to the car (properties *one* and *two*) and to the driver of a car (properties *three* to *seven*). In a self-driving car, the properties *four* to *six* must be part of an *autonomous driving system*, while the passenger determines the goal, i.e. property *seven*. The holonic enclosure restriction introduced in Sect. 5.3 stipulates that all interactions of the autonomous control system with the car are handled at the interfaces among the holons at this *level n-4* and that there must not be any lower-level interaction of the subsystem *car* with the subsystem *autonomous driving*.

The interface between the autonomous driving system and the car is relatively simple (see Fig. 10.3). It consists of (i) a periodic data stream of two state variables denoting the intended acceleration (positive: acceleration, negative: braking) and the steering angle and (ii) the stigmergic information flow of the movement of the car in its environment, observed by the sensors of the autonomous driving system. This holonic closure restriction is essential for the simplification of a hierarchy. It separates the detailed implementation of the mechanics of the car from the detailed implementation of the autonomous driving system and thus realizes the principle *Divide and Conquer* (Sect. 7.1).

Example 10.5.2: If, for economic reasons, the autonomous driving system introduced in example 10.5.1 and the computer software for the engine control system—see *level n-7* on Fig. 10.1—are executed on the same hardware platform, then temporal conflicts in the program executions of these two systems must be considered. In a time-triggered hardware platform, these temporal conflicts can be resolved at design time—when the periodic processing schedules for the task executions and the periodic communication schedules are designed. If this is the case, a correct run time environment remains free of temporal conflicts. If an event-triggered hardware platform is used, then such an *a priori* resolution of the temporal conflicts is not possible. Since it is also impossible to completely test such a system, there remains a small probability that—even in a correctly working hardware system—temporal faults will occur during the operation.

10.6 Safety-Critical Control System

Any safety-critical control system must provide a safe level of service even in the face of any sensor failure or of any hardware or software failure in one of its complex components [Bak09].

It is impossible to design a physical device that will not fail eventually, nor is it possible to develop a complex software component without a design error.

The following final Section of this book sketches an architecture for a safety-critical application that takes these insights into account. Figure 10.4 depicts the top-level components of the fault-tolerant time-triggered architecture for safety-critical control systems. Remember the following essential requirements that a safety-critical control system must meet (see Sect. 6.4):

R1: Given the objectives of the system and the constraints of the operation, the relevant state of the physical process and the environment must be periodically observed and set-point values for the actuators must be provided periodically.
R2: A *timing violation* or an *error in the calculation of a set-point value* must be promptly detected. This requires a reference for error detection.
R3: In case that a timing violation or a value error is detected a shutdown of the system to a safe state must be carried out immediately (Fig. 10.4).

There are four independent *stand-alone components* provided at the top level of this safety-critical architecture: *the primary component, the secondary component, the shut-down component* and the *decision component* These top-level components exchange data via a fault-tolerant communication system. Each one of these components is a *fault-containment unit (FCU)* such that a fault in a component—caused either by hardware or software—is contained within the component and manifests itself only as a failure at the message interface of the component to other components: either a *timing error (or a missing message)* or a *value error* in a message.

The components are executed periodically, where the time-slice of a period is determined by the dynamics of the physical process that must be controlled and the performance of the available hardware platform. A fault-tolerant global time synchronizes the start of all component executions. At the start of a time slice the primary component, the secondary component, and the shut-down component observe the environment at about the same instant (at the same tick of the global clock) and calculate—completely independently of each other—their respective results which are transmitted to the decision component at the end of the time-slice.

Fig. 10.4 Top-level
components of a safety
critical control system

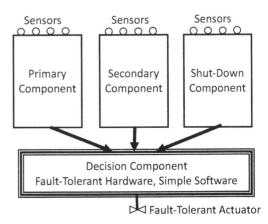

The primary component that implements requirement R1 observes the environment with its own sensors and calculates a trajectory of the set points for the actuators. In a non-fault-tolerant system, only the primary component is needed. A failure of the primary component causes an erroneous set-point value.

The secondary component provides the reference for error detection (requirement R2). The secondary component observes the environment with its own sensors and calculates a safety envelope for the set points of the actuators by a hardware-software system that is developed independently of the primary component. A safety envelope restricts the value domain of the set-points to values that do not violate any safety assertion.

The shutdown component that fulfills requirement R3 observes the environment with its independent sensors and calculates a *shutdown procedure* that will bring the process from the current state to a safe-state. In the example of an autonomous car, this means that the car is autonomously driven to stop at a safe position at the side of the road. In some applications, where a human operator is required to attend to the system, a fall back to human operation may be adequate.

The decision component which implements requirement R3, checks whether the set-point values delivered by the primary component are within the safe envelope delivered by the secondary component. If this is the case, the set points are released to the fault-tolerant actuators. If this is not the case, then the set-points calculated by the shutdown component are delivered to the fault-tolerant actuators—or the control is transferred to a human operator—and from there onwards, the set-points from the *shutdown component* are delivered to the fault-tolerant actuators until the process is in a safe state.

Example 10.6.1: If an automated subsystem in an airplane detects that it cannot handle a given situation (e.g., due to a sensor failure), then the human pilot should be requested to take control. This will only work if the automated subsystem contains error detection mechanisms to find out that it cannot handle the given situation.

The three components, the primary component, the secondary component, and the shutdown component are assumed to contain complex software, i.e., the software can contain a residual design error. The components are executed on conventional non-fault-tolerant hardware. The hardware of the three components, i.e., the computers and the sensors, should be of a different design and the software of the three components should be developed by independent development teams, such that no correlated fault exists in the three components. Let us assume that the primary component and the secondary component exhibit an MTBF in the order of 10^5 h of operation. This MTBF includes hardware and software faults. If the primary component and the secondary component are developed independently and access independent sensors, then it is *highly improbable* that both components will fail at the same instant.

The correct operation of the shut-down component is only required if there is a disagreement between the primary component and the secondary component, i.e., if a set-point calculated by the primary component is outside the safety envelope calculated by the secondary component. We assume that the shutdown component has a probability of failure on demand of around 10^{-4} [Lit93].

Following the insights of Lui Sha [Sha01, p.12]: " ... *it is the existence of a simple and reliable component that can be used to guarantee the critical properties*" — the software of the decision component must be simple. It consists of a periodic comparison and a decision based on this comparison. The decision component does not contain any concurrency. We therefore assume that after a rigorous design process and a very thorough test the software of the simple decision component is free of design errors. The hardware of the decision component is fault-tolerant to mask hardware faults.

The provided top-level architecture design allocates the requirements to independent components. R1 is satisfied by the primary component, R2 is satisfied by two nearly independent components, the secondary component and the decision component and R3 is satisfied by the shutdown component.

The proposed design is splittable in the sense that a different implementation group can implement each one of the components concurrently. There is no—*and should not be*—any communication among the groups implementing the primary component, the secondary component, and the shutdown component. The interfaces between these three components and the decision component—the periodic transmission of the trajectory of set points—consist of a periodic stream of state variables and are thus simple.

In order to increase the security and limit the impact of an intrusion, the secondary component and the shutdown component should have no connections to the outside world.

Safety-Critical Software

In the previous example of a safety-critical system we differentiated between complex software that is executed in the primary component, the secondary component and the shutdown component on non-fault-tolerant hardware and the simple software of the decision component that is executed on fault-tolerant hardware. Since the simple software of the decision component is assumed to be free of design errors and a hardware fault is masked by the fault-tolerant mechanism of the hardware, the single thread of control that runs through the decision component is expected to deliver a correct result.

This assumption about the correctness of the software in the single thread of control of the decision component is necessary, because an error in this component may cause a catastrophic failure. A replication of the decision component would bring the problem of *replica-non-determinism* into the scene [Pol96].

What are the characteristics of simple software that justify the assumption that all design errors can be eliminated before the software is put into operation? This question, which is discussed in a number of documents (e.g., [Law11]), is difficult to answer. It is easier to answer the question *When is software complex?* and to posit that if a software is *not complex* it must be *simple*. We make the tacit assumption that it is not possible to eliminate all design errors if the software is complex.

We consider software as complex if one or more of the following characteristics are present:

- *Size*: Application software size larger than 1000 lines of source code (see Sect. 6.4).
- *Cyclometric Complexity*: Cyclometric complexity of the application software greater than 10 (see Sect. 2.4).
- Use of a *general-purpose operating system*: There is empirical evidence, e.g., [Cho61, Pal11], that general-purpose operating systems contain residual design errors.

If none of these characteristics is present and a disciplined development process, an exhaustive test and a thorough certification process are followed in the development of the simple software, then we assume that the simple software is free of design errors.

10.7   Points to Remember

- The conceptual integrity of the evolving design must be defended against a multitude of *nice-to-have features of marginal utility* that are likely to lead to unanticipated feature interactions and corrupt the original conceptual integrity envisioned by the designer.
- *Essential complexity,* which is related to the *essential functionality* of a planned system, is inherent and unavoidable, given the purpose of the intended system, while *accidental complexity* is the consequence of misguided design decisions.
- The hasty acceptance of given requirements without any questioning of the rationale behind them and their relations to the essential purpose of the system introduces a *down-stream essential complexity* that cannot be mitigated any more in the following design and implementation process.

- After a number of requirement iterations that change some of the original require-ments, considering the result of trade-off studies about the efforts required to solve an identified problem by hardware or by software, the *essential hardware requirements* and the *essential software requirements* of the project can be established.
- Experience has shown that the future evolution of a successful system is inevi-table. It is thus prudent to plan for this evolution already during the architecture design phase.
- Since every computational action or every communication action can fail, mech-anisms must be in place to detect component and communication failures.
- If the system behavior is predictable under peak-load conditions, then the normal load will take of itself.
- A one-to-one relationship between an *essential requirement* and the correspond-ing top-level component that implements the associated *essential function* of the evolving system simplifies the implementation, the testing and the evolution of the essential requirements.
- In a time-triggered system the global time-schedules for the execution of the real-time tasks and the transmission of messages should contain sufficient slack to provide room for the *evolution of the functionality* of a successful system.
- The construction of a multi-level hierarchy of models to arrive at an understand-able structure of a large system requires a large effort.
- A safety critical control system requires redundancy to detect hidden design faults and hardware errors.
- We assume that it is not possible, with today's state of technology, to eliminate all design errors if the software is *complex*.

Chapter 11
Annex: Basic System Concepts

11.1 Introduction

This annex discusses some of the fundamental system concepts that have been used frequently throughout this book. For a more thorough discussion of these terms, see the book *Cyber-Physical System of Systems* [Bon16].

11.2 Entities, Time, and Systems

Entity *An entity is either a thing or a construct that is of relevance in our view of the world. A thing exists in the physical world, while a construct is a creation of the human mind, a concept.*

> **Example 11.2.1:** A stone is a *thing*, while the intention when throwing a stone is a *construct*.

(See also Sect. 2.2 and Example 2.2.2 on the difference between a *concept* and a *construct*).

Carl Popper in his *Three Worlds Model* [Pop78] distinguishes between the world of physical *things* (world one), the world of *constructs* that are the products of the human mind and are documented in libraries (world three) and the world of the human mediator (world two) that mediates between *world one* and *world three*.

Things interact either by *physical forces* or by the *exchange of matter* or *energy*. *Constructs* relate by *associations* or the exchange of *information* (see Chap. 3). The fields of logic or mathematics are examples for systems that are exclusively concerned with relations among *constructs*.

© Springer Nature Switzerland AG 2019
H. Kopetz, *Simplicity is Complex*, https://doi.org/10.1007/978-3-030-20411-2_11

In order to restrict the entities and the interactions (relations) among the entities that are at the focus of interest in a given context, we introduce the notions of a *Universe of Discourse (UoD)* and an *Interval of Discourse (IoD):*

The Universe of Discourse (UoD) comprises the set of entities and the inter-actions among the entities that are of interest when modeling the selected view of the world.

The specification of the UoD must also include the setting of the ranges of the physical parameters that characterize the considered physical environment of the things.

The Interval of Discourse (IoD) specifies the time interval that is of interest when dealing with the selected view of the world.

Property We call a quality of an entity a *property*. A physical entity, i.e. a thing, is characterized by a vast number of different properties. A property can have a *value*.

Example 11.2.2: A stone has the properties of weight, color, texture, hardness etc.. The property *weight* can have the value of *1 kg*. The property *color* can have the value *brown*.

The *value domain* of a property denotes the set of all possible values that a property can take on.

Only a small subset of the properties and interactions of an entity are required to *model* the behavior of an entity for a given purpose (see example 2.5.5 and Chap. 4).

Time The basis of our model of time is Newtonian Physics, where the arrow of time—the *time line*— flows from the past to the future. A cut of the time-line is called an *instant*. We call a happening at an instant of time an *event* and the *value of a property of an entity* that persists between two different events, a *state of the entity*. An event is thus a change of state. The observation of a state is an *event* that can capture the *value of the property.* Events cannot be observed, only the consequences of event occurrence—the new unfolding state—can be observed.

A *digital clock* is a device that produces *periodic events* (ticks of the clock) and contains a memory that denotes the *value* (a sequential number) of the state of the clock between two successive ticks that confine a *granule, i.e. the granularity,* of the clock. The granularity of a clock can only be determined if a *reference clock* with a finer granularity is available—see [Kop11, p.53 onwards]. A *timestamp of an event* by a clock is the *value* of the granule of this clock during which the event occurred.

A *global time* is an abstract notion that refers to the fact that the ticks of all local clocks of the nodes of a distributed system are *approximately* synchronized within a known *finite precision*. A *fault-tolerant global time* can be established by the peri-odic resynchronization of at least four local clocks. For a detailed discussion about global time see [Kop11, pp. 57–62].

System We are now in the position to provide a definition of our notion of a system:

A system is a collection of related entities that forms a whole.

The wholeness is a defining characteristic of a system. In many cases, a system is enclosed by a type of a *skin* that separates the system from its environment. Identifiable entities inside a system are often called *subsystems*, *parts* or *components* (see below) of the system.

The static relationship among the parts inside the system is called the *structure of the system*.

Environment of a System The entities in the UoD that are not part of the system but have the capability to interact with the system during the IoD.

A system and the environment of the system interact by *interfaces* that are located in the *skin* of the system. The skin with its interfaces thus forms a *boundary* between the system and its environment.

An entity in the UoD can only interact with a system, if *some property of the external entity* can be *observed* by a sensor or *influenced* by an actuator in the interface of the system.

The environment that is noticed by a system depends on the sensing and actuating capability of the system.

A system may be *sensitive to the progression of time,* implying that the system may react differently at different points in time, to the same pattern of input activity, and this difference is due to the progression of time [Bon16].

Example 11.2.3: A heating system with a time-controlled thermostat, where the temperature set point of the thermostat depends on the current time.

Cyber-System A system in cyber space that processes, stores, and transports information items and has the capability to interact with its environment by the exchange of information items (Itoms) via sensors and actuators.

Cyber-Physical System (CPS) A system that consists at least of two subsystems, a physical subsystem (a thing) that is being controlled (e.g., a machine) and a cyber system that observes the thing with sensors and controls the thing by sending information items to actuators that act on the thing in the physical world. A CPS may also include a human operator.

The above example 11.2.3 of a heating system is an example of a simple Cyber-Physical System.

The laws describing the behavior of the physical part of a CPS refer to physical time, while the computational progress in the cyber-system depends on the provided algorithms and on the performance of the available hardware platform.

Message A message is an atomic unit that captures the value domain and the temporal domain of a unidirectional data transport at a level of abstraction that is applicable in many diverse scenarios of machine communication and human communication. The data aspect and the timing aspect are integral parts of the message concept.

Depending on the given context, the term *message* can denote a *message instance* or a *message type*. A *message instance* refers to a particular message. A *message type*—similar to the term *data type*—informs about the *explanation* of a class of messages that have the same structure and share the same explanation.

Behavior and Service   The sequence of *message instances* produced by a system for its environment is called the behavior of the system. The intended behavior is called the *service* of the system.

Architecture   The *architecture of a system* establishes the framework for the integration and interactions of the subsystems, the *components*, of a system.

In the *time-triggered architecture (TTA)* it is assumed that a global time is available at all nodes of a distributed computer system. The control signals for the start of a computation by a component or the transmission of a message are derived periodically from the progression of this global time.

11.3   Component

The term *component* or *computational component* is widely used in the domain of information science to describe an identified subsystem of a Cyber-Physical System. In cyberspace, components interact by the exchange of *Itoms,* where the data part of an Itom is transported in a *message* across a component interface and the explanation of the Itom is contained in the component.

Let us consider a two-level hierarchy, the macro level — the level of the *whole* or the level of the *system*—and the micro level —the level of the *parts* or the level of the *subsystem* of the system. We define a computational component as follows:

> *A computational component, or component for short, is a subsystem of a large computer system that provides a specified computational service across a set of well-specified interfaces*

A (computational) component consists of a software-subsystem that is executed on a hardware platform. If the hardware platform for the execution of the component is shared with other software subsystems we call the component a *shared component*, otherwise a *stand-alone component.*

A *stand-alone component* is a *hardware-software unit* with *external communication interfaces* for the transmission of messages to other components and to the physical environment of the stand-alone component.

A *shared component* requires for its software execution a shared hardware platform. A *shared software component* is characterized by three types of interfaces: (i) an *interface to the hardware platform* (and its operating system*)* for the execution of the component software, (ii) *internal communication interfaces* to other shared components on the same hardware platform and (iii) *external communication interfaces* to the environment outside the shared hardware platform.

Ideally the structure and the protocols for the platform-internal communication should be identical to the structure and the protocols for the platform-external communication. It is then relatively easy to transfer a component from one hardware platform to another hardware platform or to make out of a *shared component* a *stand-alone* component.

> *A component is a real-time component (RT-component) if the results of the component must be available before a specified deadline.*

A RT-component must have access to sufficient hardware resources such that the execution of its software can provide a result before the specified deadline. It is the responsibility of the operating system of the shared hardware platform to ensure that sufficient hardware resources are provided to a RT-component for its execution.

A component is a recursive concept that can be applied to any level of a multi-level hierarchy. If we move one level down in a multi-level hierarchy, the *current component becomes* the *whole* of the new macro-level and the internal parts of the current component become new lower-level components. If we move one level up in the multi-level hierarchy, the *current whole* becomes a component at the new micro-level and *a new whole* comes into existence at the macro level.

Interface Component *An interface component is a component that controls a sensor or an actuator that is at the interface between cyberspace and the physical world.*

Interface components are permanently linked to the hardware of the sensor/actuator and cannot be reconfigured. The sensors/actuators produce/consume *sense data* (see Sect. 3.3). The idiosyncratic format of the sense data depends on the hardware design of the sensor/actuator system and should remain hidden inside an interface component, since it most often requires an extensive explanation (and a corresponding high mental effort) to arrive at the meaning of the sense data. Only the *refined data* that conforms to a standard representation should be visible at the external interface of an interface component.

Service Interval (SI) In a time-triggered (TT) real-time system, the duration and position of the periodic time intervals—we call it the *service intervals SI*—when the software of a RT-component must be executed on the given hardware platform is determined before the software execution, i.e., at design time. An appropriate duration of this *service interval* depends on the characteristics of the algorithms and the performance of the available hardware platform. In a TT system the deployment of *anytime algorithms* [Kop18] that are guaranteed to provide a satisficing result at the end of the service interval is a good choice in order to avoid the problems associated with the determination of the worst-case-execution-time (WCET) of an algorithm [Wil08].

> *The duration and position of the service interval (SI) on the time line are the temporal parameters that link the hardware world to the software world in a time-triggered CPS.*

11.4   Communication

Components interact by the exchange of *messages* that are transported by a communication system along a communication channel.

Structure of a Message In cyberspace a message is formed by the concatenation of three distinct fields: a *header*, a *payload*, and a *trailer*. The header contains the address of the sender, the address of the receiver and some control parameters. The header informs the communication system whereto the message has to be sent. The *payload* contains the *cargo*, i.e. a *bit-vector* that is sent from the sender to one or more receivers. The trailer consists of a cyclic-redundancy check (CRC) field to enable the receiver to check whether the payload has been mutilated during the transport. In case the receiver detects a mutilation of the message, the message is discarded.

The communication channel provides the physical means to transport a message from a sender to a receiver. The capacity of a serial communication channel, expressed in *bits per second* (or *baud rate*) depends on the physical characteristics of the channel and the electronic equipment attached to the endpoints of the channel. In the last 20 years the baud rate that can be transmitted on a twisted pair channel has been significantly increased and reaches today 1 Gigabit per second (and even more).

Send Time The total time d_{total} required to send a unidirectional message with a length *of k bits* on a channel with a *baud rate br* is given by

$$d_{total} = d_{setup} + d_{prop} + d_{trans}$$

where d_{setup} is the *setup time,* i.e. the time the sender has to wait until the channel becomes free to accept a new message, d_{prop} is the *propagation delay* of an electric wave traversing the channel (about *1 μsec* per *200 meter* channel length), and d_{trans} is the transmission time, that is the time it takes to transmit the *k bits* of the message, i.e.

$$d_{trans} = k \,/\, br$$

Event-Triggered (ET) Message In an *event-triggered communication system* the sending of a message is initiated as a consequence of the occurrence of a significant event, such as the completion of a processing task or the arrival of an interrupt from the environment. Because of an unknown *setup time* (the channels can be busy at the time of the occurrence of the significant event) the *transport latency* of an event-triggered message is unpredictable.

Time-Triggered (TT) Message In a time-triggered communication system, the sending of a message is initiated periodically at predetermined instants of a global physical time. Since the preplanned schedule for the transmission of a TT-message is guaranteed to be free of conflicts the *setup time is zero* and the *transport latency* of a time-triggered message is known *a priori*.

11.5   Failures, Errors and Faults

In this Section we take up the terminology on dependable computing community introduced in [Avi04]. Let us consider a system that is intended to provide a *service* to its environment, where an independent omniscient external observer observes the *behavior* of the system.

Failure   A *failure* occurs, when the observer detects a deviation of the actual service from the intended service.

In case of a failure, an intended message from the system is missing or a received message contains an erroneous content.

Error   An *error* is a deviation between the intended state and the actual state.

An error can only be detected if there is a *reference* available such that the result contained in a received message or the instant when a message is delivered can be compared with this reference. If the reference is not a *golden standard* (i.e., the correctness of the reference is out of question), then this comparison will only tell us that there is an error, either in the received message or the reference.

Fault   A fault is the cause of an error or of a failure.

Let us now assume that the system is composed of *components* that interact by the exchange of (system-internal) messages and that the observer can also see all of these internal messages.

Fault Containment Unit (FCU)]   A component is *a fault-containment unit (FCU)* if a fault within the component is contained by the component and manifests itself as a failure to other components solely by an erroneous or missing message. FCUs must fail independently.

Fail-Silent FCU   A fail-silent FCU is an FCU that outputs either correct messages or no messages at all.

There are a number of well-known techniques to design a fail-silent FCU, e.g. the duplication of its subsystems and the comparison of the results [Kop12, p.156]. The failure of a fail-silent FCU can only be detected in the *temporal domain,* because if a failure of a fail-silent FCU is detected in the value domain, the outputs of the fail-silent FCU are disabled.

If all components of a system are *fail-silent*, then the implementation of fault-tolerance is substantially simplified since error detection in the value domain is not required.

An FCU can fail either due to a physical (hardware) fault or due to a design fault in the hardware or in the software.

A hardware fault is *transient* if it corrupts the state stored in the hardware and has no effect on the future physical operation of the hardware. A hardware fault is *permanent* if it damages the future physical operation of the hardware.

The *mean time between failures (MTBF)* w.r.t. permanent hardware failures, i.e. physical faults, of a high-quality state of the art hardware chip with millions of transistors is better than 10^7 h, i.e., 1000 years, whereas the MTBF w.r.t. transient hardware failures is orders of magnitude smaller, depending on the physical environment of the hardware.

11.5.1 Software Errors

Jim Gray [Gra86] makes an important distinction between two types of errors in software, *Bohrbugs* and *Heisenbugs*. A Bohrbug is *logically deterministic* (see Sect. 2.2). Heisenbugs are different. Heisenbugs are time sensitive—the reproduction of a Heisenbug requires the precise duplication of a critical temporal relationship between involved concurrent processes. It follows that the cause of a Heisenbug is more difficult to find than the cause of a Bohrbug. The residual software errors in a large software system are predominantly of the Heisenbug type. In many operational situations it is impossible to decide on the basis of the appearance of a malfunction whether the cause of the malfunction is a Heisenbug or a transient hardware fault.

List of Abbreviations

API	application program interface
c-data	context data
CPS	cyber-physical system
CRC	cyclic redundancy check
DoD	degree of dependence
e-data	essential-data
FCU	fault-containment unit
HMI	human-machine interface
HO-schema	Hempel-Oppenheim schema
IIC	internal interface component
IoD	interval of discourse
ISO	international standards organization
MIM	mental interface model
MTBF	mean-time between failures
NRT	non-real-time
PMM	pragmatic mental model
RL	reporting latency
RT	real-time
SI	service interval for the execution of a component
TCS	temporal control structure
TT	time-triggered
TTA	time-triggered architecture
TTEthernet	time-triggered Ethernet
TTP	time-triggered protocol
UoD	universe of discourse
WCET	worst-case-execution time

© Springer Nature Switzerland AG 2019

H. Kopetz, *Simplicity is Complex*, https://doi.org/10.1007/978-3-030-20411-2

References

[Ahl96] Ahl, V. and Allen T.F.H. *Hierarchy Theory.* Columbia University Press. 1996.

[Ami01] Amidzic, O., H.J. Riehle, T. Fehr, C. Wienbruch & T. Elbert. *Pattern of Focal y-bursts in Chess Players.* Nature. Vol. 412. (p. 603). 2001.

[Ari15] Aristotle, Metaphysics. Aeterna Press. 2015.

[Avi04] Avizienis, A., Laprie, J.C., Randell, B., and Landwehr, C.. *Basic Concepts and -taxonomy of Dependable and Secure Computing.* IEEE Trans. on Dependable and Secure Computing. Vo.1, Issue 1, pp. 11–33 Jan. 2004.

[Avi82] Avizienis, A., *The Four Universe Information System Model for the Study of Fault-Tolerance.* Proc. of FTCS 12. Pp. 6–13. IEEE Press. 1982.

[Bak09] Bak, S. et al. *The System Level Simplex Architecture for Improved Real-Time Embedded System Safety.* 15th IEEE Real-Time and Embedded Technology and Applications Symposium. IEEE Press.. 2009.

[Bar64] Bar-Hillel, Y, Carnap, R. *An Outline of a Theory of Semantic Information.* In Bar Hillel, ed. Language and Information. Addison Wesley, pp. 221–274. 1964.

[Ben13] Joseph Bennington-Castro. *Are Humans Hardwired to Detect Snakes?* URL: https://io9.gizmodo.com/are-humans-hardwired-to-detect-snakes-1453865235 Retrieved on Aug 14, 2018.

[Bon16] Bondavalli, A. *Cyber-Physical System of Systems.* Eds. Bondavalli, A, Bouchenak S. and Kopetz H., LNCS 10099. Springer Verlag. 2016.

[Bou61] Boulding, K.E. *The Image.* Ann Arbor Paperbacks, 1961.

[Bro75] Brooks, F., *The Mythical Man Month: Essays on Software Engineering.* Addison Wesley Longman Publishing Co. 1975.

[Bro87] Brooks, F. *No Silver Bullet.* IEEE Computer, Vol. 20, No. 4, pp.10–19. IEEE Press. 1987.

[Che66] Cherry, C. *On Human Communication.* MIT Press. Second edition, 1966.

[Cho61] Chou, A, et al. *An empirical study of operating system errors.* Proc. of SIGPS 01. Pp.73–88. ACM Press. 2001.

[Cra67] Craik, K, *The Nature of Explanation.* Cambridge University Press. 1967.

[Dah66] Dahl, O.J. and Nygaard, K.. *SIMULA: an ALGOL based Simulation Language.* Comm. ACM. Vol 9. No. 9, pp. 671–678. September 1966.

[Dvo09] Dvorak, D.L.. NASA *Study on Flight Software Complexity.* Jet Propulsion Laboratory. California Institute of Technology. 2009. URL: http://citeseerx.ist.psu.edu/viewdoc/download?doi=10.1.1.711.2976&rep=rep1&type=pdf Retrieved on June 14, 2018.

[Edm00] Edmonds, B.. *Complexity and Scientific Modeling.* In: *Foundations of Science.* Springer Verlag. (pp. 379–390). 2000. URL: https://link.springer.com/content/pdf/10.1023/A:1011383422394.pdf retrieved on February 10, 2019.

[Eps08]   Epstein, S. *Intuition from the Perspective of Cognitive Experiential Self-Theory*. In: *Intuition in Judgment and Decision Making*. Lawrence Erlbaum New York. (pp. 23–38). 2008.

[Eri06]   Ericsson, A.K. *Protocol Analysis and Expert Thought: Concurrent Verbalizations of Thinking during Expert's Performance on Representative Tasks*. In: Ericsson, A.K. et al. eds. *The Cambridge Handbook of Expertise and Expert Performance*. Cambridge University Press. pp.223–242. 2006.

[Flo05]   Floridi, L. *Is Semantic Information Meaningful Data?* Philosophy and Phenomenological Research. Vol. 60. No. 2. Pp. 351–370. March 2005.

[Fur19]   Furrer, F. J. *Future Proof Software Systems. A Sustainable Evolution Strategy*. Springer Verlag. 2019.

[Gai89]   Gaines, M. *Software Fault caused Gripen Crash*. URL: https://www.flightglobal.com/pdfarchive/view/1989/1989%20-%200734.html Retrieved on June 14, 2018.

[Gib18]   Gibbens, S. *Why a Warming Arctic May Be Causing Colder U.S. Winters*. National Georgraphic, March 13, 2018.URL: https://news.nationalgeographic.com/2017/07/global-warming-arctic-colder-winters-climate-change-spd/ Retrieved on August 22, 2018.

[Gla13]   Glanzberg, M. *Truth*. Stanford Encyclopedia of Philosophy. 2013.

[Gra59]   Grasse, P.P. *La reconstruction du nid et les coordinations interindiviuelles chez Bellicositermes natalensis et Cubitermes sp. La theorie de la stigmergie*. Insectes Sociaux Vo. 6., pp. 41–83. 1959.

[Gra86].   Gray, J., *Why do Computers stop and what can be done about it?*. Technical Report 85.7. Tandem Computers. June 1985.

[Hal96]   Halford, G.S., W.H. Wilson, & S. Phillips. *Abstraction, Nature, Costs, and Benefits*. Department of Psychology, University of Queensland, 4072 Australia. 1996.

[Hei07].   Heimgärtner, R. *Cultural Differences in Human Computer Interactions: Results from two on-line Surveys*. Proc. of 10. International Symposium for Information Science. Pp. 145–157. 2007.

[Hem48]   Hempel, C.B.. Oppenheim, P. *Studies in the Logic of Explanation*. Phil. Sci 15(2). pp. 135–175. 1948.

[Hme04]   Hmelo-Silver, C.E. & M.G. Pfeffer. *Comparing Expert and Novice Understanding of a Complex System from the Perspective of Structures, Behaviors, and Functions*. Cognitive Science, Elsevier, Vol. 28. (pp. 127–138). 2004.

[Hoe16]   Hoefer, C. *Causal Determinism*. Stanford Encyclopedia on Philosophy. 2016.

[Hof17].   Hofstede, G. et al. *Cultures and Organizations—Software of the Mind*. McGraw Hill. New York. 2010.

[ISO13]   International Standards Organization (ISO). *The international language of ISO graphical symbols*. ISO, Geneva. *2013*.

[Joh83]   Johnson-Laird P.N. *Mental Models*. Cambridge University Press. 1983.

[Ken78]   Kent, W., *Data and Reality*. North Holland Publishing Company, 1978.

[Kha12]   Khan, El. and Khan, F. *A comparative Study of White Box, Black Box and Grey Box Testing Techniques*. Int. Journal of Advanced Computer Science and Applications., Vol 3 No.6, pp.12–15. 2012.

[Koe67]   Koestler, A. *The Ghost in the Machine*. Hutchinson of London. 1967.

[Kop11]   Kopetz, H.. *Real-time Systems–Design Principles for Distributed Embedded Applications*. Springer Verlag, 2011.

[Kop14]   Kopetz, H. *A conceptual model for the information transfer in Systems-of-Systems*. Proc. of ISORC. Reno Nevada, IEEE Press. 2014.

[Kop15]   Kopetz, H., *Direct versus stigmergic information flow in systems-of-systems*. SOSE Conference Proceedings 2015. IEEE Press. 2015.

[Kop16]   Kopetz, H. et al. *Emergence in Cyber-Physical Systems of Systems*. In: *Cyber-Physical System of Systems*. Eds. Bondavalli, A, Bouchenak S. and Kopetz H., LNCS 10099. Springer Verlag. 2016.

[Kop18]   Kopetz, H. *Anytime Algorithms in Time-Triggered Control Systems*. In: *Principles of Modeling*. Ed. Lohstroh M. et al. Springer LNCS 10760. Pp. 326–335. Springer Verlag. 2018.

[Kop18a] Kopetz, H., *Method for the reliable transmission of alarm messages in a distributed computer system*. US Patent 9,898.924. Granted February 20, 2018.

[Kop97] Kopetz, H.. *Real-time Systems–Design Principles for Distributed Embedded Applications*. Springer Verlag, 1997.

[Kos10] Koscher K. et al. *Experimental Security Analysis of a Modern Automobile*. Proc. of the 2010 IEEE Symposium on Security and Privacy. IEEE Press. 2010.

[Law11] Lawrence, E. *System Safety Analysis and Assessment for Airplanes*. FAA Advisory Circular No: 23.1309-1E. Federal Aviation Administration. 2011.

[Lee08] Lee, E. A. *Cyber-Physical Systems Design Challenges*. Proc. of ISORC 2008. Pp.363–369. IEEE Press. 2008.

[Lee17] Lee, E.A., *Plato and the Nerd*. MIT Press. 2017.

[Li18] Li, G. et al. *Understanding Error Propagation in Deep Learning Neural Network (DNN) Accelerators and Applications*. Proc. of SC17. Denver Colorado. ACM Press. 2017.

[Lit93] Littlewood, B., Strigini, L., *Validation of Ultra-High Dependability for Software Based Systems*. In: Randell, B., Laprie, JC., Kopetz, H., Littlewood B (eds) *Predictably Dependable Computing Systems*. Springer Verlag. 1995.

[Liu17] Liu, W. et al.. *A Survey of Deep Neural Network Architectures and their Applications*. Elsevier Neurocomputing. Vol. 234, pp. 11–26. 2017.

[Mac76] McCabe, T. *A Complexity Measure*. IEEE Trans. on Software Engineering, Vol SE-2, No 4,. pp. 308–320. 1976.

[Mat92] Maturana, H. and Valera, F., *The Tree of Knowledge: The Biological Roots of Human Understanding*. Shambhala, 1992.

[Mil56] Miller, G.A. *The Magical Number Seven, Plus or Minus Two: Some Limits on Our Capacity for Processing Information*. The Psychological Review. Vol. 63. (pp. 81–97). 1956.

[Min74] Minsky, M., *A Framework for Representing Knowledge.*, 1994. URL: https://dspace. mit.edu/bitstream/handle/1721.1/6089/AIM-306.pdf?%20sequence=2 Retrieved on August 20, 2018.

[New72] Newell, A., Simon, H., *Human Problem Solving*. Prentice Hall, 1972.

[Ogi17] Ogie, R.I., *Cyber Security Incidents on Critical Infrastructure and Industrial Networks*. Research Report University of Wollongong, Australia. URL: https://ro.uow.edu.au/cgi/view-content.cgi?referer=https://scholar.google.at/scholar?hl=de&as_sdt=0%2C5&q=Cyber+Security+Incidents+on+Critical+Infrastructure+and+Industrial+Networks.+&btnG=&httpsredir=1 &article=1217&context=smartpapers / Retrieved on October 10, 2018.

[Pal11] Palix, N. et al, Faults in Linux—Ten years later. Research Report BR-7357, INRIA France. 2010.

[Pat00] Pattee, H.H., *Causation, Control, and the Evolution of Complexity*. From: *Downward Causation: Mind, Bodies, Matter.* Editor: P.B. Anderson et al. Aarhus Unversity Press. pp. 63–77. 2000.

[Pat73] Pattee, H.H. *The Physical Basis and Origin of Hierarchical Control*. In: H.H Pattee. *Hierarchy Theory: The Challenge of Complex Systems*. Ed. George Braziller. New York. pp.73–108. 1973.

[Pia70] Piaget, J. *Structuralism*. Harper and Row. 1970.

[Pol96] Poledna, S., *Fault-Tolerant Real-Time Systems, the Problem of Replica Determinism*. Springer Verlag. 1996

[Pop78] Popper, K. *Three Worlds. The Tanner Lecture on Human Values*. Univ. of Michigan, 1978. URL: *https://tannerlectures.utah.edu/_documents/a-to-z/p/popper80.pdf Retrieved on April 18, 2018.*

[Pow95] Powell, D. *Failure Mode Assumptions and Assumption Coverage*. in: Predictably Dependable Computing Systems. Springer Verlag. 1995.

[Rai10] Rajkumar, R. Lee, I., Sha, L., Stankovic, J., *Cyber-Physical Systems: The next Computing Revolution*. 47th ACM Design Automation Conference. pp.731–736. 2010.

[Rei10] Reisberg, D. *Cognition*. W.W. Norton, New York, London. 2010.

[Ros12] Rosen, R. *Anticipatory Systems*. Springer Verlag, 2012.

[Saa08]   Saaty, T.L.. *Decision Making with the Analytic Hierarchy Process.* Int. J. of Service Sciences. Vol. 8. No. 1. Pp. 83–98. 2008.

[Sha01]   Sha, L. et al.. *Using Simplicity to control Complexity.* IEEE Software. Vol. 18. No. 4. Pp20–28. 2001.

[Sch04]   Schlindwein, S. Luis, Ison, R.. *Human knowing and perceived complexity: implications for systems practice.* Emergence: Complexity and Organization, Vol, 6(3) pp. 27–32. 2004.

[Sha93]   Shannon, C.E. *Collected Papers.* ed. by N. J. A. Sloane and & A. D. Wyner. Los Alamos, Ca: IEEE Computer Society Press. 1993.

[She17]   Shenhav, A., et al. *Toward a Rational and Mechanistic Account of Mental Effort.* Annual Review of Neuroscience. Vol. 40. Pp. 99–124. 2017.

[Sim68]   Simon, H. *The Architecture of Complexity.* In: *The Science of the Artificial.* Ed. H. Simon. MIT Press. Pp.193–229. 1969.

[Sob15]   Sober, E., *Ockham's Razors.* Cambridge University Press. 2015.

[Suh90]   Suh, N.P. *The Principles of Design.* Oxford University Press. 1990.

[Tah99]   Tahtaras, M., *Hand-Cranking, Safe and Easy.* URL: www.abarnyard.com/workshop/handcrank.htm. Retrieved on May 23, 2018.

[Tör18]   Törngren, M. and Sellgren, U. *Complexity Challenges in Development of Cyber-Physical Systems.* In: *Principles of Modeling.* Ed. Lohstroh M. et al. Springer LNCS 10760. Pp. 478–503. Springer Verlag. 2018.

[Vig62]   Vygotski, S. *Thought and Language.* MIT Press. 1962.

[Wil08]   Wilhelm, R. et al. *The Worst-Case Execution Time Problem—an Overview of Methods and a Survey of Tools.* ACM Transactions on Embedded Systems. Vol 7, No 3. 2008.

[Wri04]   Wright, G.H. *Explanation and Understanding.* Cornell University Press. 1971.

[WWW13]   World-Wide Web Consortium. Extensible Markup Language. URL: http://www.w3.org/XML/ Retrieved on April 18, 2018.

Subject Index

A

Abstraction, 2, 4, 6, 14, 16, 25, 30, 38, 51, 52, 57–59, 62, 65, 67, 78, 79, 83–88, 106, 110, 114, 117, 129

Accommodation, 13, 26, 109

Acoustic pattern, 24

Alarm shower, 91, 93, 118

Algorithms, 4, 11, 12, 25, 29, 37, 40–43, 48, 50, 52, 53, 64, 70, 71, 74, 78, 81, 82, 85, 88, 90, 93, 95, 117, 118, 129, 131

Alignment of contexts, 90, 108–110

Anytime algorithms, 42, 43, 75, 85, 117, 131

Application program interface (API), 83, 84, 88

Architectural style, 30, 101–104, 106, 110

Archival data, 28, 81, 92

Arrow of real time, 41, 52

Assimilation, 13, 26, 109

Assumption coverage, 51, 52

Atomic unit, 25, 129

Attack surface, 73, 74

B

Bit pattern, 29, 32, 96

Bizarre phenomenon, 49

Bohrbug, 134

Boundary conditions, 7, 8, 39, 48

Bounded rationality, 47

C

Causal actions, 7, 10

Causal explanation, 6–9, 16

Causal loop, 62–64, 67

Causes, 6–10, 13, 14, 16, 19, 26, 31, 49, 62, 70–72, 115, 122, 124, 133, 134

Certification, 41, 124

Chess, 7, 15, 16, 47

Client actions, 80

Clock skews, 95

Cognitive complexity, 6, 11, 12, 15–17, 46, 110

Communication, vi, 20–27, 31–35, 41, 61, 78–80, 82, 88–99, 102, 103, 106, 109, 110, 117, 118, 120, 121, 123, 125, 129–132

Compatible models, 52

Complementary models, 52

Complexity, v, vi, 1–17, 19, 30, 42, 46, 51, 74, 77–79, 81, 82, 85, 86, 102, 109, 113–115, 119, 124

Complex software, 12, 73, 75, 121, 123

Components, 12, 19, 30, 32, 57, 58, 60, 61, 66, 72, 79–84, 89, 90, 95–98, 101–106, 110, 116–118, 121–123, 125, 129–133

Composability, 60

Comprehensibility, 22

Computational complexity, 11

Concepts, 5, 19, 47, 59, 73, 77, 109, 127

Conceptualizations, 49, 51, 53, 58, 77, 85–88, 108, 116

Conceptual landscapes, 5, 6, 13–17, 22, 26, 34, 38, 44, 45, 53, 70, 102, 103, 108, 109

Confidentiality, 73

Configuration and control interface, 105

Connotation, 21, 33, 35

Consistent-ordering problem, 72

Constructs, 2, 6, 7, 16, 21, 22, 25, 35, 37, 43, 50, 127

Context data (c-data), vi, 31–33, 90, 99

Contexts, 5, 6, 20–25, 27, 29–35, 40, 42, 45, 83, 85, 86, 90, 95, 101–103, 108–110, 114, 128, 130

© Springer Nature Switzerland AG 2019
H. Kopetz, *Simplicity is Complex*, https://doi.org/10.1007/978-3-030-20411-2

Author Index

A
Ahl, V., 38, 59
Allen, T.F.H., 38, 59
Amidzic, O., 15
Aristotle, 10
Avizienis, A., 86, 133

B
Bak, S., 121
Bar-Hillel, Y., 19, 26
Bondavalli, A., 127, 129
Boulding, K.E., 6
Brooks, F., 5, 113

C
Carnap, R., 19, 26
Cherry, C., 19
Chou, A., 124
Craik, K., 44

D
Dahl, O.J., 4
Dvorak, D.L., 11, 73, 113

E
Edmonds, B., 15
Elbert, T., 15
Epstein, S., 13
Ericsson, A.K., 47

F
Fehr, T., 15

Floridi, L., 19
Furrer, F.J., vi, 116

G
Gaines, M., 108
Gibbens, S., 49
Glanzberg, M., 22
Grasse, P.P., 33
Gray, J., 134

H
Halford, G.S., 14
Heimgärtner, R., 109
Hempel, C.B., 7
Hmelo-Silver, C.E., 35
Hoefer, C., 8
Hofstede, G., 109

I
Ison, R., 11

J
Johnson-Laird, P.N., 38, 44
Joseph Bennington-Castro, 46

K
Kent, W., 19
Khan, E., 39
Khan, F., 39
Koestler, A., 57, 66

© Springer Nature Switzerland AG 2019
H. Kopetz, *Simplicity is Complex*, https://doi.org/10.1007/978-3-030-20411-2

Printed in the United States
By Bookmasters